Mapping Human History

DISCOVERING
THE PAST
THROUGH OUR
GENES

Steve Olson

HOUGHTON MIFFLIN COMPANY
BOSTON · NEW YORK
2002

For information about permission to reproduce selections
from this book, write to Permissions, Houghton Mifflin Company,
215 Park Avenue South, New York, New York 10003.

Visit our Web site: www.houghtonmifflinbooks.com.

Library of Congress Cataloging-in-Publication Data
Olson, Steve, date.
Mapping human history : discovering the past through
our genes / Steve Olson.
p. cm.
Includes bibliographical references and index.
ISBN 0-618-09157-2
1. Human population genetics. 2. Physical anthropology
and history. 3. Human genetics — Variation.
4. Human beings — Migrations. I. Title.
QH455 .O474 2002
599.9 — dc21   2001051880

Printed in the United States of America

Book design by Robert Overholtzer

QUM 10 9 8 7 6 5 4 3 2 1

The excerpt from *Siddhartha* by Hermann Hesse, copyright © 1951
by New Directions Publishing Corp., is reprinted by permission
of New Directions Publishing Corp.

Parts of this book, in different form, first appeared in
"The Genetic Archaeology of Race," *Atlantic Monthly,* April 2001.

For Lynn

Contents

IV. EUROPE

V. THE AMERICAS

VI. THE WORLD

Mapping Human History

The Human Pageant

ANYONE WALKING along the sidewalk of a large city can't help but be struck by the incredible variety of human beings — tall and short, fat and lean, hairy and hairless. Some have skin the color of heavy cream; others are as dark as bittersweet chocolate. The shapes of people's faces, the colors of hair and eyes, the contours of eyes, noses, and lips are marvelously unique. Partly we're attuned to these differences because we use them to identify people we know. But our diversity is not an illusion. Human beings really are one of the most physically varied species on earth.

People tend to use these different appearances to draw conclusions about the ancestry of others. In the Americas and Europe, if you have dark skin, people assume that you have African ancestors, though dark-skinned peoples also come from southern India, Australia, and parts of southeastern Asia. Individuals whose eyelids have an extra fold of skin are assumed to be from Asian families, though not all Asians have an epicanthic fold, as it is called, and non-Asians scattered from southern Africa to the Americas do. A prominent nose, deep-set eyes, and light skin identify a person as having European ancestors, though people with similar features, if slightly darker skin, have lived for thousands of years in India, Polynesia, northern Japan, and North and South America.

Our propensity to sort people into categories has, over the course of history, contributed to immense human suffering. Entire groups have been condemned to slaughter or slavery because of the color of their skin or the shape of their eyes. Even today, almost all the armed conflicts in the world take place not between nations but between groups separated by differences that often are interpreted in biological terms. In places like Rwanda, the Balkans, Indonesia, and the Middle East — to take only the most recent examples — groups of people have fought and killed each other with an almost tribal savagery.

Elsewhere, tension between physically distinguishable groups often lurks just beneath the surface of our everyday lives. In the United States, more than three-quarters of African Americans and European Americans alike judge relations between the two groups to be only fair or poor. In other parts of the world, arguments over immigration and the assimilation of minorities have sparked deadly riots and public protests. In 1900 the African-American scholar W. E. B. Du Bois said, "The problem of the twentieth century is the problem of the color line." At the dawn of a new century, the issue of race looms larger than ever.

Groups of people vary in appearance because their ancestors had different biological histories. But just how different were those histories? What if the distinctive appearances of human groups are a historical accident, a biological joke, no more substantial than masks at a costume party?

The story that I'm going to tell in this book begins in Africa, somewhere in the high grasslands and wooded slopes of present-day Ethiopia, Kenya, and Tanzania. More than 100,000 years ago, a small group of humans lived in this region. They gathered nuts, fruits, and seeds from the countryside. They hunted gazelles and hares and scavenged meals from the carcasses of animals killed by other predators. They looked much like people today, with high foreheads, sharp chins, and light, graceful bodies.

Many other humans lived elsewhere in Africa, Europe, and Asia at that time, but they were noticeably different from us. Their forehead slanted backward toward a low, bunched skull, and a heavy visorlike ridge protruded from just above the eyes. These people would never have passed what anthropologists call the subway test — that is, if you gave them a shave, dressed them in modern clothes, and seated them on a subway car, they could never hope to blend in. The other passengers would inevitably look at them and say, "That's a mighty strange-looking human being."

These so-called archaic humans vastly outnumbered the small group of anatomically modern humans living in eastern Africa. More than 1 million archaic humans may have been scattered across the Old World, while the population of modern humans may have dropped as low as a few tens of thousands at times.

Today all the archaic humans are gone. The last one died many thousands of years ago. Though they successfully occupied the earth for hundreds of thousands of years, they were an evolutionary dead end.

Every single one of the 6 billion people on the planet today is descended from the small group of anatomically modern humans who once lived in eastern Africa. The group occasionally came close to extinction, but it never died out completely, and eventually it began to expand. By about 100,000 years ago, modern humans had moved north along the Nile Valley and across the Sinai Peninsula into the Middle East. More than 60,000 years ago they made their way along the coastlines of India and southeastern Asia and sailed to Australia. About 40,000 years ago, modern humans moved from northeastern Africa into Europe and from southeastern Asia into eastern Asia. Finally, sometime more than 10,000 years ago, they made their way along a wide plain joining Siberia and Alaska and spread down the length of North and South America.

Wherever modern humans encountered their archaic counterparts in Africa, Asia, and Europe, the old-style humans eventually disappeared. Many questions surround these encounters. According to the available evidence, the two groups interbred very little, if at all. This conclusion seems totally at odds with our knowledge of human history, which shows that groups of people have eagerly interbred at almost every opportunity. Perhaps modern and archaic humans were so different genetically that they were not able to produce offspring. Or maybe they did have children, but hybrids were not accepted into bands of modern humans, so the archaics' physical characteristics could not enter modern populations. We just don't know.

How modern humans replaced their predecessors is also a mystery. Archaeological evidence indicates that in some places bands of modern and archaic humans lived near each other for thousands of years. Yet no signs of warfare have been found. The cave paintings of Europe, some of which were being drawn just as the archaic Neandertal people were dying out on the continent, show plenty of violence against animals but none against other humans.

Until a few years ago, much of what I've written in the last few paragraphs was not known. Before then, the only way to learn about our ancient ancestors was through the scattered bones and stone tools they left behind. Such evidence is maddeningly scarce: of the several billion modern humans who lived before the invention of agriculture, scientists have found the fossilized remains of perhaps a few hundred. And the scarceness of the evidence means that each fragment of bone and chipped rock bears a heavy burden of speculation.

But bones and stones are not the only record of our past. Each of us carries around another record in almost every cell of our bodies. Human DNA, the long, complex molecule that transmits genetic information from one generation to the next, bears the indelible imprint of human history. Our DNA records the evolution of an African ape that began walking on two legs more than 4 million years ago. It documents the emergence of modern humans on the savannas of eastern Africa about 7,500 generations ago. It chronicles the diversification of modern humans into the "races" and "ethnic groups" that we recognize today.

Geneticists are just beginning to read the story written in our DNA, but already they have discovered a saga of immense grandeur. They now can trace the movement of modern humans out of Africa into the rest of the world. They are beginning to piece together when and how groups of people acquired their distinctive physical appearances and what those appearances mean. They are learning how groups mixed and diverged over time. They are discovering the immense gulf that separates what actually happened in the past from the stories we tell ourselves about the past.

This is a book about history, but it is a history with a very real presence in our everyday lives. Most of us think of ourselves as Hispanic or Chinese, white or black, Nigerian or Norwegian, or some combination of categories. For some people these labels mean very little, while for others they are the single most important aspect of their identity. But no matter what importance individuals attribute to these labels, there is no denying their continued power in modern societies. Many people continue to think that human groups have fundamental biological differences. They believe that outward variations in skin color, facial features, or body shape reflect much more consequential differences of character, temperament, or intelligence. Even when two groups are physically indistinguishable, people tend to cite genetics as the source of group differences. They say that the aggressiveness or religiosity or inventiveness of a group could not possibly be learned — it must be something in the genes.

Genetics research is demonstrating otherwise. Human groups are too closely related to differ in any but the most superficial ways. The genetic study of our past is revealing that the cultural differences between groups could not have biological origins. Those differences must result instead from the experiences individuals have had.

Much of this new knowledge about our past is emerging from an unexpected source. Research into the genetic causes of disease is simultaneously revealing the histories of groups and of individuals. Biomedical researchers now realize that they need to learn about the genetic differences between individuals to understand why some people get sick and others remain healthy. But these genetic differences are a consequence of human history — of the matings of particular men and women over thousands of years. By studying those differences, researchers are reconstructing the history of our species.

I first became interested in this story in 1987, when studies of human genetic variation revealed that everyone on earth is descended from one woman who lived in Africa around 150,000 years ago. Over the next few years I worked as a writer on several projects, mostly at the National Academy of Sciences, that used DNA comparisons to trace the evolutionary relationships of species, for instance, or to understand the origins of a disease. Gradually I realized that human DNA is a virtually limitless repository not just of biomedical but of historical information, a sort of molecular parchment on which an account of our species has been written. I began to compile articles and books on the subject, and my stacks of documents grew.

The story contained in our DNA is rich, complex, and multilayered. In this book I'll look at five broad regions of the world — Africa, the Middle East, Asia and Australia, Europe, and the Americas — with a final chapter on Hawaii. In each section I'll trace the history of modern humans in that part of the world from their first appearance to the present. At times the story leads in unexpected directions — to the origins and diversification of language, for example, or to the experiences of particular groups such as the Jews or Han Chinese. But my central concern will always be the same: to explore what our genetic history tells us about ourselves and about our past, present, and future as a species.

The study of human genetic differences is one of the most contentious areas of modern science; to some people, such investigations are too dangerous to pursue. In their view, genetics research is a modern Pandora's box that threatens to reinforce stereotypes and limit human potential.

I see things differently. We are on the verge of conquering scourges that have plagued human beings throughout history — hunger, disease, mental illness. Admittedly, our new knowledge will pose great risks. We will learn about individual susceptibility to diseases, which could change how we view our lives and our relations to others. We will know much more about our history on this planet, both as individuals and as members of a group. We have to find ways to interpret, handle, and protect this powerful new information.

But we are not a species that shies away from knowledge. On the contrary, for thousands of years human beings have sought innovative ways to do things and to organize their lives. Genetics research has presented us with an opportunity to relieve the world of great suffering. Turning our backs on this opportunity would run counter to our conception of ourselves.

Many of the medical applications of genetics still lie in the future; we have time to think about the dilemmas that those applications will pose. But the data needed to reconstruct the genetic history of humanity are available right now; we need to figure out what they mean. If the frequency of a particular genetic marker is high in one group and low in another, are those groups fundamentally different? To what extent can Asians, Spaniards, Polynesians, American Indians, or any other group be defined genetically? Can genetics be used to assign individuals to groups?

We needn't fear such questions. On the contrary, with the proper safeguards in place, genetic information can be a powerful liberating force. In the past we never really understood why groups of people had distinct appearances. Bigots therefore could read into those physical differences anything they wanted. Genetics research is now about to end our long misadventure with the idea of race. We now know that groups overlap genetically to such a degree that humanity cannot be divided into clear categories. We know that human behavior is tremendously malleable under the influence of different social settings. The story written in our DNA is one of great promise, not peril.

Besides, it's one of the best stories you'll ever hear. It has adventure, conflict, triumph, and sex — lots of sex. It ranges from jungles to deserts to icy plains, across generations and thousands of years. It's the story of us, from our humble origins on the savannas of Africa to a position of unprecedented mastery over our own future.

I

Africa

1

The End of Evolution

The African Origins of Modern Humans

> I am an African. I owe my being to the hills and the valleys,
> the mountains and the glades, the rivers, the deserts, the
> trees, the flowers, the seas and the ever-changing seasons
> that define the face of our native land.
>
> — Thabo Mbeki, from a speech delivered upon the adoption
>   of the constitution of the Republic of South Africa,
>   May 8, 1996

THE PAST AUTUMN has been the rainiest season in southern Africa in more than a century, and the scrublands of northeastern Botswana are bursting with life. Hornbills and shrikes glide among the acacia trees. The bush is rich with flowers and seed. The leopard that lives in this area, which no one has seen for months, left paw prints last night a hundred yards from our camp.

About a dozen Bushmen are moving languidly through the underbrush. They are following the tracks of a small antelope that passed this way a couple of hours ago, but they are not really serious about hunting. A young man named Xoma (that's how he spells his name in English, though in fact it begins with a complicated click sound that's very difficult to pronounce) spots a familiar vine. With a few quick jabs of his digging stick, he unearths a plump tuber the size of an orange. He hands the prize to a nearby woman, who stashes it in the leather kaross slung over her shoulder, then hurries off to join the other men for a smoke.

The lives of these people, who call themselves the Ju/'hoansi and are also known as the !Kung San, have changed dramatically in recent decades. Xoma and his family now live in a permanent house made of wood and tin rather than the thatch huts that the Ju/'hoansi used to construct when they established a new hunting camp. At school the Ju/'hoansi children learn the national language of Botswana, not the complex click-based language their ancestors spoke. They wear shirts and slacks, not the traditional leather clothes made from the animals they hunted. Young men of Xoma's age often leave the bush to work elsewhere in Botswana or in neighboring South Africa.

But for a few weeks each year, members of Xoma's village move back into the bush to live in the old ways. They forage for roots with weighted digging sticks. They hunt with bows and arrows and cook the spitted game over crackling fires. They talk and joke for hours while carving ostrich shell beads or playing an impenetrable game in which they move stones among indentations scooped from the ground. Xoma is learning to be a healer. At night, when the Bushmen gather around the fire to sing and clap the rhythms of ancient songs, he dances with uncertain steps behind his mentor, learning to achieve the trance state that will connect him with the spirit world.

Though they are fast becoming part of a cash economy, many of the Bushmen who live in this part of Botswana still obtain some of their food by hunting and gathering in the land surrounding their villages. But disputes with neighboring ranchers and farmers are common, and the allure of a more modern life is powerful. Whether the tradition of hunting and gathering will survive for much longer remains to be seen.

The Bushmen are the original people of southern Africa. (The equivalent words "Bushmen" and "San" both have derogatory connotations, but no other terms for this group of people are available, and many of them prefer "Bushmen" because of its association with the land.) Their ancestors have lived here for tens of thousands of years, perhaps for more than 100,000 years. Over that time the Bushmen developed a way of living in harmony with each other and with the land. They took what they needed for the present while ensuring that enough remained for the future. They built elaborate social networks through marriages, alliances, and trade. They left many thousands of paintings on rock walls scattered across southern Africa.

But over the last few millennia, other groups have steadily en-croached on their homelands. Somewhat more than 1,000 years ago, groups of farmers and herders who were taller and had darker skin began to push into southern Africa from the north. Gradually the Bushmen either mixed with the invaders or retreated into less produc-tive lands. Then, in the 1600s and 1700s, Dutch farmers began to spread north from the Cape of Good Hope. Although the Bushmen and their neighbors fought desperately to stop the settlers, gradually the Europeans prevailed.

Throughout the history of their contact with others, the Bushmen have been the objects of a virulent racism. Other Africans have often treated them as vagrants and thieves. (One meaning of "San" is "un-trustworthy.") Many European farmers, on the other hand, simply de-cided that the Bushmen were not human. A late-nineteenth-century tally from German South-West Africa lists the animals shot by settlers and policemen over the previous year. At the top of the list, under the heading "mammals," is "female Bushmen: 400."

Denying the humanity of other people has always been a way to jus-tify oppressing and exterminating them, and science has a long, sad history of contributing to these atrocities. Well into the twentieth cen-tury, anthropologists were speculating that Africans, Asians, and Eu-ropeans had evolved from different kinds of primates. The clear impli-cation was that these groups belonged to different species, one of which was more highly evolved than the others.

But one obvious problem has always plagued this idea. If two ani-mals belong to different species, they rarely are able to interbreed. Yet whatever other limitations human beings have, the inability to inter-breed has never been one of them. Southern Africa today is a genetic hodgepodge of groups descended from the Bushmen and their pasto-ral cousins the Khoi Khoi, from neighboring farmers and herders, and from European and Asian immigrants. The Xhosa, the group to which Nelson Mandela, Thabo Mbeki, and many other South African leaders belong, obviously has some Bushman ancestry. The "Cape Coloureds" are the descendants of European pioneers, Asian immigrants, and the indigenous people of southern Africa. Many European South Africans have African ancestors from the early years of European settlement, when different groups extensively interbred. One of the great ironies of the apartheid era in South Africa, when people were divided into rigid racial categories, is that few countries have such a rich legacy of genetic mixing.

Anyone who lives in Africa can immediately recognize a group of Bushmen. They are small and wiry. Their skin color ranges from reddish brown to almost yellow. Their hair grows in tightly wound tufts and is so brittle that it naturally breaks off. With prominent cheekbones and delicate features, they are a handsome people by today's standards.

Why are the Bushmen so distinctive in appearance? For that matter, what makes any group of humans recognizable? What accounts for the distinguishing features we use to categorize people?

Where the Bushmen live certainly has a big influence on their appearance. The faces of elderly Bushmen are deeply lined from constant exposure to the sun. Physical activity and diets rich in vegetables have given most of them a lean, sinewy physique.

But the underlying reasons for the Bushmen's similarities to one another require a closer look. Under a microscope, the cells in the top layer of their skin are indistinguishable from those of people anywhere else in the world. But deeper in their skin, beneath the transparent uppermost layers, are the cells known as melanocytes, which give skin its color. In Bushmen these cells are darker than those of Europeans and Asians because they contain larger amounts of the pigment eumelanin. On the other hand, the melanocytes of the Bushmen are lighter than the heavily pigmented cells of Africans whose ancestors lived closer to the equator.

Beneath the melanocytes, the differences between the Bushmen and other people again fade away. Every other type of cell in their bodies looks no different from the corresponding cells in other people. In that respect, the differences between the Bushmen and anyone else on earth are truly skin deep.

But skin color is just one attribute. What about the Bushmen's small bodies, pointed chins, and hooded, almost Asian, eyes? To find the origins of these differences, we have to look into the nucleus, the small compartment that exists inside almost every human cell. Floating in the nucleus, in a warm bath of nutrients and enzymes, are forty-six structures called chromosomes. There are twenty-three pairs of chromosomes in humans, numbered 1 to 22 in order from longest to shortest. The twenty-third pair consists of an X chromosome and a Y chromosome in males or two X's in females. (The vast majority of people have the usual complement of twenty-three pairs, but a few have extra chromosomes — such as individuals with Down syndrome, who have an extra chromosome 21.)

The pairwise organization of chromosomes reflects the mysterious dualism of sex. One chromosome in each pair is descended from a chromosome in the father's sperm cell; the other is descended from a chromosome in the mother's egg. In that respect, each pair is like a married couple, bound until death. The pairs even engage in their own form of sex. When an adult organism begins to create new sperm or egg cells, the chromosome pairs delicately intertwine and exchange pieces in a process known as recombination. The result is two hybrid chromosomes, as if a husband and wife had exchanged arms and legs. These hybrids are separated, packaged in new egg or sperm cells, and sent on their way to begin the process anew.

The odd couple are the X and Y chromosomes. Egg cells always contain an X chromosome. Sperm cells contain either an X or a Y. Fathers are therefore responsible for the sex of their offspring, though it is largely a matter of chance whether a Y-bearing or an X-bearing sperm swims up the fallopian tube, finds a fertile egg, and is the first to breach the egg's inner sanctum.

Except for the X and Y, the two members of each chromosome pair are almost identical. (This is where the husband and wife analogy breaks down.) They have to be, or the cells of the body would not work properly. For example, when the members of a chromosome pair exchange pieces during recombination, the chromosomes have to match up, like partners on a dance floor. If the chromosomes are incompatible, the dance cannot proceed, and the process of reproduction grinds to a halt.

When people think about chromosomes, they often recall a picture from a school biology textbook. At a certain point in the life of a cell, the chromosomes scrunch up into stubby cigar-shaped objects. If they are then exposed to a chemical called Giemsa stain, bands appear around the chromosomes like the stripes on a croquet mallet.

Except for people with rare chromosomal abnormalities, these banding patterns are essentially the same for people anywhere in the world. When male white settlers mated with female Bushmen in the eighteenth century, their corresponding chromosomes lined up perfectly. In the search for the origins of the Bushmen's distinctive attributes, the chromosomal banding patterns offer no clues.

But chromosomal banding patterns do differ from species to species. We often hear, for example, that human beings and chimpanzees are remarkably alike genetically. And, when stained and compared, some human and chimp chromosomes in fact cannot be visually distinguished from one another. A careful comparison turns up the telltale differences, however. Chimps have twenty-four pairs of chromosomes, not twenty-three, and some of the banding patterns are subtly different. On nine of the chromosomes, certain segments are flipped in humans compared with chimps. On other chromosomes, extra material is tacked onto the ends, or some is missing. These differences embody the evolutionary distance between our species. Our lineages have been separated for so long that the structure of our chromosomes has diverged.

If the banding patterns of the chromosomes tell us nothing about the differences between the Bushmen and other people, then we must look deeper. Each chromosome contains a single strand of deoxyribonucleic acid, or DNA. DNA has achieved an almost iconic status in our society. Biotech companies build double-helix staircases for their headquarters. Glossy magazine illustrations show the molecule twisting away into a dimly seen future. Shampoos trumpet their DNA content, as if the inclusion of anything from a plant or animal must be good for our hair. (Counterexamples are easy to find. A major constituent of DNA, guanine, got its name from guano, from which the molecule was first isolated.)

The problem with icons is that we tend not to think deeply about them, which is unfortunate in the case of DNA, because it really is one of nature's most amazing creations. First of all, molecules of DNA can be incredibly long. If the DNA in the forty-six chromosomes of a single human cell were stretched out, it would extend from one side of a kitchen table to the other — six feet altogether. It seems impossible that so much material could be packaged inside an object smaller than a dust mote. The secret is DNA's thinness. If the six feet of DNA on the kitchen table were enlarged until it extended from New York to Los Angeles, the molecule would still be no wider than a pencil.

Even more astonishing than the length of DNA is how much information it can hold. The core of a DNA molecule consists of four simple building blocks known as nucleotides — adenine, thymine, cytosine, and guanine, abbreviated A, T, C, and G — strung together in a chain. For example, a particular section of DNA on human chromosome 2 consists of the following nucleotides: ATACTGGTGCTGAAT. But that's just 15 nucleotides. The twenty-three chromosomes in each human sperm or egg cell contain about 3 billion nucleotides altogether — 6,000 times as many nucleotides as there are letters in this book. Electronics engineers often congratulate themselves on the amount of data they can cram into a semiconductor chip. They have a long way to go to catch up with the information density of DNA.

The string of nucleotides in DNA looks like gibberish to us. But that's because we don't speak the language. To the cell, the messages embodied in DNA are the wisdom of the ages. Each of us inherited our DNA from our biological mother and father, who in turn got their DNA from our grandmothers and grandfathers. The first creatures who could be called human inherited their DNA from creatures that could not be called human. The first mammals got their DNA from their reptilian ancestors. And so it goes, back through time, to the first single-celled organism that began using DNA to transmit genetic information. DNA is our link to every other creature that has ever lived on this planet.

If an earnest graduate student took copies of chromosome 2 from two people and began comparing the chromosomes' nucleotide sequences, the student would find the two sequences to be almost identical. But about once in every 1,000 nucleotides, on average, the two sequences would differ. One person might have an A at that point, while the other has a G. Or a few nucleotides might be added, deleted, or transposed in one person but not the other.

Here at last is the origin of the genetic differences between individuals and groups. All humans everywhere in the world have exactly the same set of genes. But many of the genes come in slightly different versions. These differences in the DNA sequences of our genes lie at the base of our physical uniqueness. They create the color of our skin, eyes, and hair. They generate the shape of our skull, the distribution of hair on our head, and the overall contours of our bodies. They influence our likelihood of getting particular diseases. They are the biological foundation on which we build our lives.

Much of biomedical research in the twenty-first century will revolve around the genetic differences among people. Biotechnology companies are already studying how DNA varies from person to person to figure out how these variations contribute to disease. Soon drugs will be available that are designed to work in concert with each person's unique DNA. Eventually, biomedical researchers will figure out how to change specific nucleotides in the cells of the body and in the egg and sperm cells that create new people. At that point, humans will be able to take evolution into their own hands and, for better or worse, will be able to determine the genetic future of our species.

But this book is not about biomedical research. It's about what DNA tells us about our past. Vast amounts of historical information have emerged from genetics research in the past few years, and much more is on the way. It's as if biologists had discovered a book written in code by observers from another planet that recounts the convoluted history of a particularly unusual species on the earth, the species we know as *Homo sapiens.*

Two thousand miles northeast of Botswana, the equator first makes contact with the continent of Africa just north of the city of Kismaayo, Somalia. From the rocky shoreline the equator sweeps up a scrub-covered hill, through the banana plantations around the Jubba River, and onto the barren flatlands that separate Somalia from Kenya. About 400 miles from the ocean the equator climbs the northern flank of Mount Kenya, passing just beneath the mountain's glistening snowfields, and then dives into the Great Rift Valley, with its eroded streambeds and hot, flat plains. On the western edge of Kenya the equator plunges into Lake Victoria, passes almost directly over the airport at Entebbe, Uganda, and then parallels the Katonga River to Lake George. On the lake's west side, the equator rises again to cross the Ruwenzori Range — the fabled Mountains of the Moon — and then quickly drops into the basin of the Congo River, passing through hundreds of miles of thick, sparsely populated rain forest. Finally, more than 2,000 miles from the Indian Ocean, it cuts through the tropical lowlands of Gabon and leaves the western side of the continent 30 miles south of Libreville, a city founded by French naval officers in 1849 as a refuge for freed slaves.

The four most important events in human evolution probably all occurred in eastern Africa within some 500 miles of the equator's passage across the continent.

- About 6 million years ago, a population of African apes split into two distinct species. One of those species would lead, through many intermediary species, to human beings. The other would lead, through a different set of intermediary species, to modern chimpanzees.

- More than 4 million years ago, one of the species on the evolutionary path to humans began spending most of its time on two feet. This upright stance seems to have set in motion a profound evolutionary trend. This new species could use its hands in new ways — to manipulate objects, for example, or throw stones to scare away predators. It could peer over the underbrush and ponder what it was seeing. No one knows exactly what led to dramatically larger brains in the species leading toward humans, but bipedality may have been a key factor. So important was this transition to verticality that paleoanthropologists — scientists who study the fossil remains of humans and their ancestors — place these apes in a distinct category. In the system used for designating species — in which the first, or genus, name denotes a group of similar species and the second name indicates the exact species — the apes that stood upright are placed in the genus *Australopithecus*.

- Around 2 million years ago, a species of large and particularly brainy biped began to alter natural objects for use as tools. Members of this species knocked pebbles together to create sharp-edged stones that they could use to butcher animals killed by other carnivores. They used bones as hammers and anvils to break apart other bones. This transition to toolmaking marked another milestone in our history. This species was the first to deserve the genus name *Homo*.

- Finally, sometime between 100,000 and 200,000 years ago, a new group within the genus *Homo* appeared. It was different from any previous group of humans: less heavily built, more mobile, with a cognitive flexibility unknown before. This is the group of humans from which we are all descended.

This chronology tends to imply that a fairly straight line led from *Australopithecus* through early *Homo* to modern humans. This is the view portrayed, for example, in the illustrated parade of ancestors that accompanies so many books and articles on human evolution. A modern human, almost always male, leads the parade, marching resolutely toward the edge of the page. He is followed by something resembling a caveman, then a bipedal ape, and finally a shambling, foolish-looking chimpanzee. The picture seems to suggest that we are the end result of a preordained process, the inevitable goal of evolution. It reinforces our belief that we are at the apex of a great pyramid of life, with all other living and extinct organisms arrayed below us.

But this picture of human evolution is wrong — or at least so incomplete as to be seriously misleading. Human evolution has not been a straightforward slog from lower to higher. It's been a maze of dead ends, unexpected detours, and sudden changes of direction. Many of the fossils that we have assumed belong to our ancestors probably represent failed evolutionary experiments, lineages of different kinds of humans that did not survive. In the end, we are the product of a relentless winnowing process, a trial by extinction.

Today, just a single human species lives on the earth. But for portions of the history of both *Homo* and *Australopithecus*, several anatomically distinct human species lived on the planet, often in the same general region. About 1.8 million years ago, as many as four distinct groups of humans and australopiths may have shared the same homeland in eastern Africa. One was a heavily built species known as *Australopithecus boisei* (many paleoanthropologists place the more robust australopiths in a separate genus, *Paranthropus*). The other three were distinctive populations of *Homo*. Only one of these populations was ancestral to modern humans, and in the absence of a detailed fossil record, we don't know which one it was.

Colin Groves was one of the first biologists to propose that different species of humans lived in a single region at the same time. A physical anthropologist at the Australian National University in Canberra, Groves has led a life of high academic adventure. He has traveled to more places and seen more kinds of animals than almost anyone on earth. In a newspaper column he writes for the *Canberra Times,* he regularly debates the views of creationists, psychics, and other pseudoscientists. A special passion is the fight against the poaching of rhinos, tigers, bears, musk deer, gibbons, and other animals whose populations have been decimated by the demand for parts of their bodies mistakenly believed to have medicinal properties.

Groves has put human evolution in context. He has studied the evolution of tigers in southeastern Asia, wolves in the Canadian Arctic, buffalo in North America, rhinoceroses and elephants in Africa, and many other mammals. Why, he asks, should the evolution of humans be different from that of any other species? Granted, humans now have an advanced culture, but this is a relatively late development in our history. And it is not clear that the attributes of early humans altered underlying evolutionary processes. "That human beings are animals like any other is perhaps the most important message of the Darwinian revolution," he says.

When Groves looks at the evolution of large mammals, he sees a distinct pattern. Species do not change en masse from one form into another. Rather, they tend to remain largely unchanged for long periods of time. Then, quite suddenly, a small group of animals with characteristics distinct enough to constitute a new species buds off from the larger group. If the new species has an advantage over the parent group, it may take over areas occupied by that group. In fact, if the advantage is great enough, the previous species may go extinct, setting the stage for the process to begin again.

This is exactly what Groves sees in the human fossil record. "Speciation events have been the prime mechanism in human evolution," he says. "Gradual change within continuous lineages has been a minor factor." In other words, the australopiths did not gradually change into the earliest *Homo* species, according to Groves. Rather, the first *Homo* species budded off from one of the existing *Australopithecus* species and gradually outcompeted its predecessors. By a million years ago, all the australopiths were gone.

Groves discovered something else in his study of mammalian evolution. Biologists have long thought that speciation occurs most often on the edge of a species' range, where a small group of animals could more easily become isolated from the larger group, perhaps by moving beyond a mountain range or river. Once isolated, this small group would evolve independently of the larger group. Over time the small group could become so different that it would no longer interbreed with its parent group, gaining evolutionary independence.

That's what Groves expected to see when he began his investigation of mammalian evolution — species spinning off new species like drops of water flung from a twirled umbrella. But he repeatedly found something quite different. New species were not forming on the edges of a range; they were forming right in the middle. Take lions, for example. Four subspecies of lions have existed in historical times. The centrally located subspecies occupies eastern and central Africa. The other three subspecies lived in southern Africa, northern Africa, and western Asia (two of the peripheral subspecies today survive only in zoos or game reserves, and the third is extinct). According to the traditional view, the newer species would be those on the periphery. But the situation is reversed. The central subspecies, with its larger brain and more complex social organization, is clearly the newest.

Exactly the same pattern can be seen in human evolution, Groves says. "All of the major events in human evolution have occurred in eastern Africa — there doesn't seem to be any question about it." According to this view, it's not an accident of geology or history that most of the important human fossils have been found in eastern Africa. For our species, that's where the evolutionary action has been.

The notion that new species can form in the center of an existing species' range is counterintuitive, but an accumulating body of evidence supports the idea. The center of the range is where resources tend to be greatest, and thus population densities are high. Dense populations produce rich stores of genetic diversity on which evolution can act. From these genetic hot spots, new variations can emerge.

The savannas of eastern Africa certainly fit this description. The highlands of Kenya and Tanzania are a biological paradise. More large mammals live in these areas than anywhere else on earth. It is a corner of the world overflowing with life.

But one major objection must be addressed: how can a new species become biologically isolated right in the middle of an existing species' range? According to the usual understanding of speciation, the new species would begin interbreeding with the existing one and quickly lose its distinct identity.

This quandary has forced biologists to look more carefully at what they call reproductive isolating mechanisms. These mechanisms have a job much like that of the chaperones accompanying high school seniors on an overnight graduation trip: to prevent successful matings. Some reproductive isolating mechanisms depend on behavior. For example, the mating rituals of two species may be so different that sexual overtures end in a welter of mixed signals and inappropriate responses. Others are more visual. Primates, in particular, with their highly acute vision, often can identify suitable mating partners at a glance.

Geography can act as a mechanism to prevent mating, even in the very center of a species' range. For example, our closest primate relatives tend to live in clumps, even where their overall populations are greatest. In the forests of equatorial Africa one can travel great distances without coming across any gorillas, then hit an area where gorillas are common, then again travel for miles with no gorillas present. Under such conditions, populations are largely isolated and can begin to diverge.

Other reproductive isolating mechanisms work at the molecular level. For example, if two individuals from distinct populations do succeed in mating, compounds on the surface of an egg can detect sperm from a different species and keep them from penetrating the cell wall. Or, if a sperm does enter an egg, the chromosomes may be incompatible and fail to produce an embryo. Finally, even if a hybrid organism is born, it may be sterile, as in the case of mules, born from the union of horses and donkeys.

The existence of many kinds of reproductive isolating mechanisms supports the observation that the creation of new species is far from unusual. On the contrary, the biological world seems to have evolved to ensure the continual production of new species. This is how nature experiments with life. If a new species is better adapted to an environment than an existing species, it will prosper. If a new or existing species is deficient, it often is thrown on the bone heap of history.

Human evolution clearly displays this harsh reality. Once the first species in the genus *Homo* appeared, it began to spin off new varieties of *Homo*. At least one of these groups did something that the australopiths had never done. Its members spread out of Africa into Asia and Europe. More than a million years ago a species known as *Homo erectus* was living in modern-day Indonesia. In Europe, a population that we today call Neandertals survived until about 30,000 years ago. In Africa, species with names such as *Homo heidelbergensis* and *Homo ergaster* strutted across the stage during various portions of the past 2 million years.

Then, sometime less than 200,000 years ago, a distinct group of humans appeared, almost certainly in eastern Africa. Initially this group must have been very small, occupying an area that may have been no larger than present-day Israel. Though its existence may at times have been precarious, it survived.

The oldest known fossil skull thought to belong to a member of this group was discovered by Richard Leakey in 1967 on the shore of the Omo River in Ethiopia. Dated to about 130,000 years ago, the reconstructed skull has a steeply rising forehead, a relatively flat face, and much smaller brow ridges than previous members of *Homo* had.

Compared with earlier skulls, this one seems hauntingly familiar. An observer peering into its empty eye sockets cannot escape the impression that this face, despite its antiquity, is that of an ancestor.

The human fossil record can be interpreted in other ways, and science cannot yet say exactly which path led to anatomically modern humans. Some paleoanthropologists, for example, adhere to a much more gradualist view. They maintain that dividing past humans into strictly separated species is a mistake. Instead, they say, humans have remained a single species ever since the evolution of *Homo* almost 2 million years ago. They demote the various kinds of humans that spread throughout Asia and Europe to subspecies or races, which continued to breed with each other and to evolve until all became fully modern.

This hypothesis is known as multiregionalism, and it is the culmination of an old tradition in paleoanthropology. It maintains that the differences we see today between various human groups predate the evolution of anatomically modern humans. In other words, Africans descend in part from the archaic *Homo sapiens* who lived in Africa, Asians descend in part from the *Homo erectus* who lived in Asia, and Europeans descend in part from the Neandertals of Europe.

Some paleoanthropologists have always had doubts about this view. But in 1987 it came under fire from an entirely new quarter. A group of molecular biologists at the University of California, Berkeley, decided to study human evolution by comparing the DNA sequences of people around the world. Their results are as astonishing today as when they were first published.

The Berkeley geneticists — Allan Wilson (who died of leukemia in 1991), Rebecca Cann, who is now at the University of Hawaii at Manoa, and Mark Stoneking, now at the Max Planck Institute for Evolutionary Anthropology in Leipzig, Germany — did not study the DNA in chromosomes. Rather, they focused on a tiny object called a mitochondrion, hundreds of which are scattered through most cells. Mitochondria have the important function of breaking down complex compounds into a simple and highly energetic molecule — a sort of cellular battery pack — that cells use to run many different chemical reactions. Oddly enough, mitochondria are almost certainly the descendants of bacteria that began to live inside other single-celled organisms more than a billion years ago. They have hitched along for the ride ever since, providing plant and animal cells with energy in return for a comfortable place to live.

Because of their independent origin, mitochondria have their own DNA — a circular loop about 16,500 nucleotides long. Each person has trillions of mitochondria in his or her body, and the DNA sequence in each of these mitochondria is, with rare exceptions, identical. But the DNA sequence of my mitochondria probably differs from the DNA sequence of yours, which is where the story becomes interesting.

We all receive our mitochondria from our mothers. Sperm cells have just one thing on their mind — delivering their package of chromosomes into the egg. Sperm cells do have a few mitochondria, but they are tossed away like worn-out flowers in the process of fertilization. Therefore, only egg cells contribute mitochondria to the next generation.

Because mitochondria are maternally transmitted, all the human mitochondrial DNA sequences that exist in the world today are descended from the mitochondrial DNA of a single woman. The first time I heard this statement I thought it highly implausible. All 6 billion people on the planet descended from a single ancestor? Yet this is one of those wonderful scientific conclusions that is not only true but *has* to be true.

About 3 billion female humans are alive right now, all of whom received their mitochondrial DNA from their mothers. (The 3 billion men on earth also received their mitochondrial DNA from their mothers, but they can be ignored in this analysis.) Now think about the generation of women one step removed from today's women, the generation that includes the mothers of every female alive today. Not all of the women of that generation had daughters. Some had only sons; some had no children at all. So the mitochondrial DNA of today's 3 billion females must be derived from a subset of the mitochondria lineages that existed in the previous generation. (The reason that more females are alive today than in the previous generation is that some women had more than one daughter.)

The same conclusions can be drawn about the grandmothers of all the women alive today. Fewer grandmothers than mothers contributed mitochondrial DNA to today's women, so the number of mitochondrial lineages from two generations ago represented in people today must be smaller than the number of such lineages from one generation ago. This reduction in the number of extant mitochondrial lineages continues with each previous generation — from the billions, into the millions, down to the thousands, and finally into the hundreds, tens, and single digits.

Eventually, the number of extant mitochondrial lineages must shrink to two — two women from whom everyone in the world today is descended. Perhaps those two female lineages extended for some time back into the past, with each daughter getting her mitochondrial DNA from a separate mother. But mathematically this process cannot continue forever, because the number of extant lineages can only decline with each previous generation — it cannot increase. When the inevitable decline occurs, the women at the roots of the two separate lineages would have to be sisters.

Their mother was the woman who produced all the mitochondrial DNA on the planet today. She is sometimes called mitochondrial Eve, but the name is profoundly misleading. It implies that she was the only woman alive on the planet at the time. In fact, many other humans lived at the same time as Eve. All those humans received their mitochondrial DNA from a different (perhaps nonhuman) female who lived in their past. But Eve's is the only mitochondrial DNA from her time that has survived. All the rest has gone extinct.

Geneticists call the process I've been describing *coalescence,* in which a DNA sequence in many people can be traced back to a sequence in just one person. The term is somewhat confusing, because coalescence proceeds as one goes backward in time. But it is a very powerful way of thinking about our DNA, because it can be applied not just to our mitochondrial DNA but to the billions of nucleotides in our chromosomes as well. Consider the Y chromosome, for example. Men pass most of their Y chromosome on to their sons in the same way that women pass their mitochondrial DNA to their children, intact and unadulterated. If a man doesn't have sons, his Y chromosome dies with him. This means that the same winnowing process that characterizes mitochondrial lineages applies to the Y. All 3 billion of the Y chromosomes that exist today coalesce in the Y of a single man who lived sometime in the past. But this "Adam" for the Y chromosome didn't necessarily know Eve. He probably didn't live at the same time. The path taken by his Y was independent of the path taken by Eve's mitochondrial DNA.

The situation is somewhat more complicated for the DNA in the other chromosomes. The recombination process that takes place between pairs of chromosomes shuffles the genetic contributions of mothers and fathers. Still, the coalescence process does apply to particular chunks of the chromosomes. Each chromosomal segment in the 6 billion people on earth today — for instance, the ten nucleotides at the end of all the chromosome 6s in the world, to take a random example — derives from the chromosome of a single person who lived sometime in the past. In fact, geneticists have made a rough guess of the number of people who contributed all of the DNA existing in the human population today. Using various simplifying assumptions, it turns out that something like 86,000 individuals, of whom mitochondrial Eve and the Adam of the Y chromosome were two, are the sources of all the human DNA in existence today.

As I've mentioned, the mitochondrial DNA sequences of today's people differ from person to person. No one now has exactly the same mitochondrial sequence as did Eve, even though we all get our mitochondrial DNA from her. The sequences of mitochondrial DNA therefore must have changed over time as they were passed from mother to children. The next chapter discusses the source of these changes. What's important to know now is that the number of differences in all of the world's existing mitochondrial DNA sequences reflects the length of time that has passed since coalescence. In essence, each new generation adds variation to DNA. As a result, the total amount of variation in a DNA sequence measures the number of generations that have passed since the coalescence of that sequence. In looking at the differences in people's mitochondrial DNA, the University of California geneticists were trying to figure out when mitochondrial Eve lived.

What they discovered is that Eve lived about 200,000 years ago. More recent analyses of mitochondrial DNA, using additional data, have produced a somewhat more recent coalescence: around 150,000 years ago. Similar calculations for the Y produce a coalescence at about the same time (though other calculations point toward a more recent Y coalescence).

Geneticists do not yet know for sure why human mitochondrial DNA coalesces about 150,000 years ago. But they do know that a coalescence is much more likely to occur when a population is small. In a small population, a few people are the ancestors of a large number of descendants, so their genes are passed through what is called a genetic bottleneck.

This process may sound familiar by now. A genetic bottleneck is what would be expected when the first anatomically modern humans appeared. A small group of humans could have been isolated for an extended period. Over time this group could have evolved a new set of characteristics that set it apart from other humans. This view fits extremely well with the genetic evidence, which indicates that our ancestors went through a bottleneck between 100,000 and 200,000 years ago. At that point, perhaps 20,000 people constituted the population from whom we are all descended.

Not everyone agrees with this picture, and the evolution of anatomically modern humans was undoubtedly more complex than my description might imply. A group of archaic humans did not disappear into a valley in Africa and emerge several thousand years later thoroughly modernized. Early modern humans might have been subdivided in some way. Modern humans and archaic humans might have interbred, though clear genetic evidence of such mixing has not been found. Much more genetic data and new fossil discoveries are needed to arrive at airtight conclusions.

But no matter what the complexities, the genetic evidence available today points to a straightforward conclusion. According to our DNA, every person now alive is descended from a relatively small group of Africans who lived between 100,000 and 200,000 years ago.

When the Berkeley geneticists first published their findings, the idea that all people share a common origin caught the public's attention. A particularly notorious illustration on the cover of *Newsweek* showed a topless black Eve offering an apple to a black Adam. But what became known as the "Out of Africa" hypothesis also was roundly attacked. Some anthropologists found holes in the analyses. Multiregionalists raised a host of objections. By the early 1990s, the attacks had placed the hypothesis in some disrepute.

The idea's rehabilitation has received less attention, but it is nevertheless solid. As geneticists began looking at other segments of DNA, they found similar signs of coalescence as they had found in the mitochondria. At the same time, the holes in the original genetic analyses were filled and new analyses confirmed the original findings.

Equally important, other scientists began to look at the fossil and archaeological evidence in a new way. A single origin fits the pattern observed in other animals, for which multiregional evolution is unknown. And an African origin for modern humans would explain why anatomically modern fossils found on that continent are older than those found elsewhere. In the most thorough computerized analysis of fossilized skulls ever done, Marta Lahr, a paleoanthropologist from Sao Paulo, Brazil, who is now at Cambridge University, found little support for the multiregional hypothesis. "The bulk of the chronological and genetic data indicate a single origin of all modern humans in Africa," she concluded.

In a way, the details of how modernity in humans arose are not important. What must count as one of the most profound biological insights of all time is the recognition of our remarkable genetic similarity. About 7,500 generations have passed since our ancestors lived on the savannas of eastern Africa. In evolutionary terms, that's the blink of an eye. The chimpanzees living on a single hillside in Africa have more than twice as much variety in their mitochondrial DNA as do all the 6 billion people living on the earth, because today's species of chimpanzees have been in existence much longer than have modern humans.

With the appearance of modern humans, the large-scale evolution of our species essentially ceased. Since then, humans have expanded as a single and relatively well mixed population. Our physical characteristics have changed slightly. We have become somewhat smaller and lighter, especially when the invention of agriculture changed the demands made on our bodies. But our basic body plan was set more than 100,000 years ago. Since then, we have been in a period of evolutionary stasis.

Everyone on earth today is equally distant from the early modern humans of eastern Africa. In that respect, no one group, including the Bushmen, is more closely related to our ancestors than any other. The same number of generations separates Australians, Canadians, and Ethiopians from early modern humans.

But different groups have different genetic histories, and those histories hold clues to the past. That's why the Bushmen are especially intriguing. According to the archaeological and genetic evidence, one of the earliest expansions of modern humans was into southern Africa. The ancestors of the Bushmen were living there when Europe and Asia

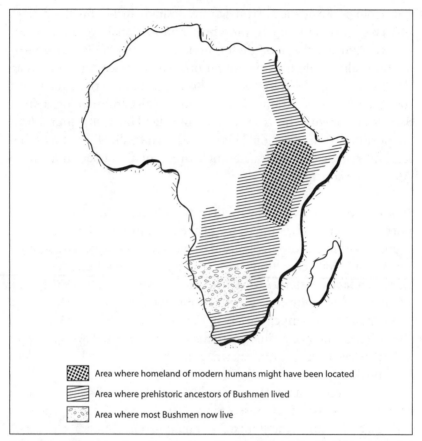

Area where homeland of modern humans might have been located

Area where prehistoric ancestors of Bushmen lived

Area where most Bushmen now live

After the appearance of modern humans, the ancestors of the Bushmen expanded to occupy large portions of eastern and southern Africa. Today the Bushmen live almost exclusively in arid regions of southwestern Africa.

were still populated with archaic humans and when the Americas were completely devoid of humans. What's more, in their southern African cul-de-sac, the ancestors of the Bushmen appear to have been relatively isolated. Certainly they mixed with people from farther north, and undoubtedly they changed over the millennia. But of all the histories described in this book, that of the Bushmen is the most straightforward.

Though it is pure speculation, maybe the Bushmen, because of their antiquity and isolation, offer a window onto the early history of modern humans. Perhaps they retain some of the characteristics of our early modern ancestors. In that case, if groups of humans have undergone any important evolutionary changes since the appearance of modern humans in Africa, maybe evidence of those changes can be found by comparison with the Bushmen.

I've spent only a few days in the backcountry of southern Africa; I can't claim to know the Bushmen well. But I've spent enough time with them to confirm what people who have spent years with the Bushmen say — they are not fundamentally different from anyone else. I've seen the affection the Bushmen lavish on their children, who grow up at the center of extended families of aunts, uncles, and cousins. I've seen their sardonic sense of humor, which they turn as easily on themselves as on the visitors who come to ask them questions. I've seen the intelligence they apply in their hunting and gathering, and the aesthetic sense they bring to their music and dance. The Bushmen seemed to me no different from other people, though of course they live in very different circumstances. So if they are at all representative of early modern humans, then humans haven't changed much over the past 150,000 years. And that's exactly the conclusion geneticists have been drawing from our DNA.

2

Individuals and Groups

The Divergence of Modern Humans

Tous parents, tous différents.

— Title of a 1992 exhibit at the Musée de l'Homme in Paris

SIBONGILE MAKHAYA'S ORDEALS WITH RACE began before she was born. Her father was an Englishman working in Johannesburg. Her mother was a South African — part Zulu, part Sotho. Under the apartheid-era Immorality Act, her mother could have been jailed for bearing the child of a European man. So she fled Johannesburg, away from Sibongile's father, away from apartheid, back to Lesotho and the safety of her family. That's where Sibongile, with skin the color of wet sand and brown wavy hair, was born.

"I grew up between two worlds," says Sibongile, an assertive woman in her early twenties who was working as a freelance video producer in Johannesburg when I met her. "My mother and I moved back to Johannesburg when I was five and lived in Soweto, because it was the only place we could live. But it wasn't an easy place for someone like me to grow up.

"I was raised as a black. I learned the languages that were being spoken around me and by my mother and her family. But when I walked outside, or in the convent school where my mother enrolled me, I couldn't be black, because the blacks treated me as coloured, or as white. The other black girls in school would grab me and choke me and say, 'How did you learn how to speak our language? You're not black.' I never could be black enough or white enough.

"Then, when kids used to say that I was white, I beat them until they bled. That's how much I hated my white blood. I hated that part of myself.

"I've had to learn how to accept myself. I've had to learn about my other culture. When I look in the mirror now, I try to see both sides of who I am."

Sibongile's life uniquely reflects both the tortured past and the guarded promise of modern South Africa. Yet her story, with minor variations, could be repeated almost anywhere in the world. Wherever people use physical appearance to judge the character of others, the mindlessness of bigotry is close at hand. People are no longer individuals with unique attributes; they're seen only as representatives of types.

But the physical differences among groups of modern humans could not always have existed. When modern humans first appeared in eastern Africa, there were so few of them that they could easily have constituted a single panmictic group. Panmixia means that people choose their mates from throughout a population rather than from distinct subgroups within it. With this continual mixing of people, physical attributes blend. As a result, subgroups tend not to form. Individuals differ from each other in a panmictic group just as they do in any human population. But the differences are all between individuals, not between groups.

If modern humans were still panmictic, there would be no obvious physical distinctions among groups. That is, if human mating were entirely random — if a person were just as likely to find a mate on the other side of the world as across the street — humans from different regions would look more or less the same. But the human species is far from panmictic today. A Korean is much more likely to have children with another Korean, a Frenchman with a Frenchwoman, an Inuit with another Inuit. As a result, the genetic variations responsible for our different appearances tend to remain localized within groups, and groups retain some measure of physical distinctiveness.

For a long time, people have tried to use the physical differences among groups to divide human beings into discrete categories or "races." They have sorted humans into three races, five races, thirty races, even thousands of "microraces." Many schemes have been proposed; none has worked. There are too many exceptions, too much overlap among groups. Humans just don't sort neatly into biological categories, despite all the attempts of human societies to create and enforce such distinctions. Meanwhile, the word "race" has become so burdened with misconceptions, so weighed down by social baggage, that it serves no useful purpose. The sooner its use can be eliminated, the better.

As we'll see in this chapter, when we think in terms of race, we are inevitably misinterpreting an extremely complex biological reality. In fact, all people are closely related through innumerable lines of descent that defeat any attempt to divide humans into races. For example, someone thinking about Sibongile's English father and South African mother might assume that her parents are about as genetically separate as two people could be. Yet a closer look at our history reveals the many ways in which they are genetically linked. Human beings really are *tous parents, tous différents* — all related, all different.

If everyone inherits his or her DNA from the same people — if we all receive our mitochondrial DNA from mitochondrial Eve, and all men inherit their Y chromosomes from a single man, and everyone gets the rest of their DNA from 86,000 other people who lived sometime in the past — then an obvious question is why everyone's DNA is not exactly the same. Why isn't everyone on earth as similar as identical twins?

I consider this the single most important question in all of human biology. It asks how the different appearances of human groups arose, why genetic diseases occur, and what role our genes play in determining who we are.

Actually, the basic question is even broader. It's the fundamental question at the heart of biological evolution: how the fantastic diversity of living things arose from the first simple organism that used DNA to reproduce.

Whenever a cell divides, it has to copy its DNA so that each daughter cell has a complete set of chromosomes. This process is amazingly accurate. The molecular machinery that reproduces DNA can continue for millions of nucleotides without making a mistake.

But the process is not perfect — if it were, after all, the earth might still be populated only by copies of that first primordial organism. Consider the 16,500 or so nucleotides in mitochondrial DNA. When a woman's mitochondria are copied and packaged into an egg cell, the sequence of mitochondrial nucleotides in the egg is almost always the same as in all of the mother's other cells. But occasionally a mistake occurs. Maybe one nucleotide is switched. Maybe instead of an A, a particular location has a G, or a T, or a C. Perhaps a stray bit of genetic material is added at a particular spot, or some is deleted. Any such mistake in the copying of DNA is known as a mutation.

Mutations can occur in other ways. Parts of our DNA can move around and alter existing sequences. A stray bit of radiation or a nasty chemical can garble a section of DNA. The cell has repair mechanisms that attempt to fix such damage. But these mechanisms don't always leave the DNA the same as before. The result is a new mutation.

These mutations are the key to reconstructing our genetic history. Suppose that mitochondrial Eve had two daughters, one of whom happened to have a single mutation in her mitochondrial DNA. Then all of the women alive today who are descended from that daughter would share her mutation, whereas all of the women who are descended from Eve's other daughter would not. In other words, mitochondrial Eve would have produced two separate mitochondrial lineages. Each of these distinct mitochondrial sequences is called a *haplotype*, and a group of related haplotypes descended from a single ancestral haplotype is called a *haplogroup*.*

Haplotypes and haplogroups will crop up often in this book. They are like pedigrees that geneticists use to learn who is related to whom. To take an oversimplified example, say that one of Eve's daughters — the one with the mutation in her mitochondria — moved to southern Africa and was the ultimate source of all the Bushmen's mitochondria, whereas the other daughter stayed in eastern Africa and was the mitochondrial ancestor of everyone else in the world. Then all the Bushmen would have the same mutation Eve's daughter had, while everyone else would not. In fact, that's how geneticists know that the ancestors of the Bushmen were one of the earliest distinct populations to appear after the evolution of modern humans. They have the oldest mitochondrial DNA haplotypes found anywhere in the world.

The loop of DNA carried in mitochondria is so small that mutations in it are rare. But the DNA sequences of our chromosomes are 400,000 times longer than those of our mitochondria. Because there is so much DNA in our cells, every act of human procreation produces

---

* On rare occasions a mutation is reversed by a subsequent mutation. But by the time the second mutation occurs, other mutations have usually altered nearby DNA sequences, and these later mutations can be used to identify the original haplotype.

unique mutations in chromosomal DNA. (The nucleotide sequences of a child's chromosomal DNA typically differ at about one hundred locations from the sequences of the mother and father.) Therefore, each newborn child is genetically unique; even identical twins have a few unique mutations that occur after the fertilized egg divides in two.

When people grow up, the mutations that they inherited from their parents are reproduced in sperm or egg cells — along with the new set of mutations that will define the genetic uniqueness of the next generation. In other words, each generation superimposes new mutations on the DNA it inherits. The result is an elaborate genealogy, an intricately branching tree of genetic alterations.

One critically important characteristic of mutations is that they occur only in individuals. A particular mutation can't appear in a whole group of people at once, like a sudden preference for flamboyant hats. It occurs in a single cell and then spreads to the next generation of cells as the first one divides. A given mutation therefore has only two ways of appearing in more than one person. The same mutation can happen independently in two people, a rare but occasional event. Or the mutation can be passed from an ancestor to more than one descendant.

You now know enough population genetics to understand everything else in this book. Admittedly, the patterns generated by hundreds of mutations occurring in millions of people interbreeding over thousands of generations can be fantastically complex. But there is an order to that complexity. Individual mutations are like drops of water falling on the surface of a pond. The waves from each drop spread out in concentric circles, mixing and blending with the waves from other drops. With hard work — and a large enough computer — the locations of the drops can be reconstructed from the observed pattern of waves.

That's what geneticists are doing when they use genetic variations to reconstruct human history. They are trying to figure out when and where specific mutations occurred and how those mutations spread in human populations. When the evolutionary lineages leading to humans and to chimpanzees split, mutations occurred in the human lineage that did not occur in the chimp lineage. When modern humans appeared on the savannas of eastern Africa, particular mutations distinguished them from their archaic neighbors. When the ancestors of today's Native Americans migrated across the Bering land bridge from Asia more than 10,000 years ago, they carried unique genetic differences that can still be found in their descendants and not in other humans.

The mutations in humans living today reveal where our ancestors lived, with whom they mated, and how individuals and groups are related. Mutations are the words in which the story of our genetic history is written.

Few people know this story better than Himla Soodyall, a human geneticist who splits her time between the South African Institute for Medical Research and the University of the Witwatersrand. Soodyall works in a graceful old building on the edge of downtown Johannesburg. In the hall outside her office, the laughter of technicians and the clink of glassware emanate from labs conducting prenatal and paternity tests — the daily grind of genetics that keeps medical institutes afloat.

A short, energetic woman with a quick smile, Soodyall is the second of three children of an Indian family that has lived in South Africa for four generations. She grew up in Durban, attending schools and a college that admitted only Indians. When she went to Johannesburg for a master's program in biotechnology, the only place she could reside under the laws of apartheid was an exclusively Indian area twenty-five miles from the university — so she spent most of her graduate years living in a nearby hotel.

"Graduate school was a difficult time for me," she recalls. She was interested in doing scientific research, but as an Indian woman in a world that was largely male and largely white, she didn't seem to fit in. Still, she earned a fellowship to Pennsylvania State University, and there she met other students who were not part of the mainstream, students who were struggling to find a place for themselves. "I had to learn that no one was superior to anyone else," she says. "When I lost my sense of being a second-class citizen, that's when I became the person I am today."

Soodyall's specialty is the mutational history of mitochondrial DNA. She knows that stretch of genetic material as well as most of us know the layout of a childhood home. She has studied the mitochondrial DNA of people from all over the world, poring over the variations in their genetic sequences. And from those variations Soodyall has painstakingly helped to piece together how mitochondrial DNA sequences have changed over time.

Soodyall is one of the researchers who have determined that the Bushmen of southern Africa harbor some of the world's oldest mitochondrial haplotypes. The Ju/'hoansi, in particular, have unique mutations that must have occurred not long after the time of mitochondrial Eve. These mutations have been passed from generation to generation among the Bushmen for more than 100,000 years. It's clear proof, says Soodyall, that modern humans appeared first in Africa, since that is where the oldest mutations exist.

But modern humans did not stay in Africa. Sometime less than 100,000 years ago, a small group left northeastern Africa and moved into Eurasia. This exodus left a clear signal in the mitochondria of people alive today, because the humans who left Africa did not carry all of the mitochondrial mutations that had occurred since Eve. They carried just some of those mutations. If humans were a deck of playing cards — like the people ruled by the Queen of Hearts in *Alice's Adventures in Wonderland* — it's as if just the sevens, eights, and nines left Africa. As a result, the mitochondrial DNA of Africans today remains somewhat more diverse than that of non-Africans.

"We believe that the base of our genetic diversity was established here in Africa," says Soodyall. "At some point a group of people who carried just a subset of that diversity left Africa, and they gave rise to all of the populations that we see today outside of Africa. Every time we add more data to the model, it makes more sense."

A couple of caveats need to be kept in mind when interpreting mitochondrial DNA sequences. Larger populations tend to create greater diversity in DNA, and until the last few millennia, more humans lived in Africa than anywhere else. Surely some of the extra diversity seen in the DNA of Africans today arises from their greater numbers over all of human history. Also, studies of mitochondria sample just a small part of our DNA. Geneticists tend to use mitochondrial DNA as an indicator for what has happened with the rest of our DNA, since it tells an exceptionally clear story. But different parts of the genome tell somewhat different stories, and these different accounts are still being compared and synthesized.

Nevertheless, the message emerging from our DNA is clear. Everyone alive today is either an African or a descendant of Africans. People on different continents do not have distinct evolutionary histories. Modern humans evolved first and then spread out to occupy the world.

Soodyall echoes a refrain common among researchers who have taken a hard look at our genes. "These data have the potential to abolish racism," she says. "Race is purely circumstantial. It establishes a social hierarchy that people can use to exploit others. But that hierarchy has no basis in biology."

If modern humans walked from Africa to Eurasia to begin their colonization of the world, their descendants could just as easily have walked back to Africa — and many did. As we'll see, the genetic evidence reveals continued mixing of Africans and non-Africans throughout the history of modern humans.

But the first departures of modern humans from Africa marked an important milestone. When human groups divide into widely separated subgroups, people are more likely to mate within their subgroup. At that point mutations occurring in one person's sperm or egg cells can spread within the subgroup, passing from a single ancestor to many descendants. But those mutations are less likely to spread beyond the subgroup, because matings occur less frequently across group boundaries. As a result, subgroups tend to accrue their own distinctive store of mutations. Genetically, the subgroups begin to diverge.

The fate of a mutation depends in part on how it affects the body. In that respect, mutations have much in common with lottery tickets. First of all, the vast majority of mutations are worthless, because they have absolutely no effect on our bodies. (For that matter, the vast majority of our DNA has absolutely no effect on our bodies. Less than 2 percent serves any known function; some of the rest may have regulatory or structural functions, but its purpose remains obscure.)

Most mutations are neutral for another reason. DNA has a fair amount of redundancy built into it. Even if a mutation occurs in a functional region of DNA, the mutation may not change the underlying function. In that case a mutation can persist from generation to generation as a harmless variant.

Not all mutations are benign. The second most common category consists of those that harm an organism. In humans, most of these mutations are so damaging that they result in miscarriages. At least 20 percent of all conceptions end in miscarriages, and many of these result from genetic mutations that are fatal to a fetus. Other severe mutations do not exert their effects until after birth. Approximately 5 percent of children are born with serious physical probelms, many of which are caused by genetic mutations. These abnormalities impose an enormous psychological and economic toll on individuals and their families. One reason that governments invest so much money in biomedical research, which can seem exceedingly abstract, is to understand and overcome these genetic disorders.

The third category of mutations, by far the rarest and most mysterious, are those that benefit an organism by making it more likely to survive and reproduce in a particular environment. If one organism leaves more offspring than another, its mutations, including its beneficial mutations, will become more common in the next generation. As Charles Darwin realized, this is the process that has filled the world with such an amazing array of living things.

The spread of beneficial mutations has played an important role in human history. Consider the relationship between skin color and climate. When people live in equatorial regions, dark skin is a great advantage because it is less susceptible to damage by the sun's ultraviolet rays. Dark skin helps prevent both skin cancer and severe sunburn, which can lead to serious infection. West African blacks who have a genetic mutation that disrupts the production of melanin usually develop skin cancer at a young age. If they die before they reproduce, that mutation cannot be passed on to the next generation, and the trait disappears from the population until the next such mutation occurs.

However, in parts of the world where sunlight is less intense, dark skin can be a liability. Our bodies use ultraviolet light that penetrates the skin to synthesize vitamin D. If vitamin D is insufficient, people's bones cannot grow properly, which can lead to the painful and disfiguring disease of rickets. Near the equator enough ultraviolet light penetrates even very dark skin to make vitamin D, but at higher latitudes pigmented skin blocks out too much light. Today the vitamin D added to milk, cereals, and other foods has helped to eliminate rickets almost completely in northern climates. Still, dark-skinned children in areas of low sunlight are at risk of rickets if they don't receive vitamin D in their diet.

One common consequence of childhood rickets is a narrowing of the pelvis. Women who have suffered from the disease are at much greater risk of dying during childbirth. As women with dark skin moved north from Africa, their incidence of death during childbirth probably increased because of their genetic susceptibility to rickets.

But say that one woman had a mutation that made her skin lighter. Such a woman would be less likely to suffer from rickets, because her light skin would allow the production of more vitamin D. She would therefore be more likely to survive childbirth. Her children who inherited the mutation from her would also have this advantage, and the mutation would spread.

Evolutionary biologists have come up with similar hypotheses for many human features. Maybe the hooded eyes of Asians protected their ancestors from the glare of the sun reflected from Siberian snow-fields. Perhaps Pygmies are small so that they don't have as much body mass to cool in the sweltering rain forests of equatorial Africa. Maybe frizzy hair and broad noses likewise helped Africans and aboriginal Australians beat the heat.

These are plausible-sounding stories, but they can't be the whole truth. Take skin color. It's true that skin tends to be lighter the farther people live from the equator. But exceptions to this rule are common. The aboriginal inhabitants of Tasmania, who were exterminated by European settlers, had very dark skin, yet Tasmania is as far from the equator as is Boston or northern Japan. Many Eskimos have darker skin than one would expect, given where they live. Some American Indian groups live in extremely sunny climates, but their skin is about the same color as that of Indians who live in the rainy Pacific Northwest.

Past migrations explain some discrepancies. The Tasmanians certainly migrated to their island home from more equatorial regions. But they lived in Tasmania for more than 8,000 years, which should have been plenty of time for their skin to lighten.

Other factors also must be in play. One such factor, surprisingly enough, is pure chance. Any time a group is isolated genetically, some mutations become more or less common entirely at random. For example, the detailed shape of our ears makes no difference to our hearing, but that shape is influenced by DNA. Therefore, average ear shape might change in one group because of random mutations while staying more or less the same in another group.

But most of our physical features don't change entirely at random.

They are influenced by a force that is especially strong among humans — our cultural preferences for particular appearances. People choose their mates (or have their mates chosen for them) for many reasons, but a particularly powerful, if inevitably superficial, reason is physical appearance. In any society, attractive individuals are unlikely to lack for suitors and therefore are more likely to have offspring. In this way, certain physical characteristics tend to be selected for in a population.

These sought-after characteristics may vary greatly from culture to culture. Say that a small group of modern humans expanding north from southeastern Asia developed a preference for the epicanthic fold. People would choose each other for this feature, and the DNA causing it would become more common. As a small group grew in size, these whimsical human preferences would become frozen in place.

Natural selection, genetic happenstance, and cultural preferences can all cause the physical characteristics of groups to diverge. But a powerful force is always pulling in the opposite direction: what Noel Coward called "the urge to merge." No group of modern humans has ever been reproductively isolated for long. Even island populations such as those of Australia, Hawaii, and Madagascar have absorbed new people, and therefore new mutations. This mixing of populations has helped to ensure that no group of humans has ever become very distinct genetically from any other.

The distribution of mitochondrial haplotypes offers a perfect illustration of this genetic connectedness. For example, the Bushmen have the oldest mitochondrial haplotypes in the world, but not all Bushmen have them. Some have haplotypes that are more common among other African populations, or even among European or Asian groups. Over time, all human groups exchange mates with their neighbors, so a mutation occurring in one part of the world can spread widely.

This mixing of DNA is evident in even a casual inspection of the world's peoples. A person traveling due east from Madrid to Beijing (both cities at about 40 degrees north latitude) would encounter Italians, Greeks, Turks, Armenians, Uzbeks, Tajiks, Kazakhs, Uighurs, Mongolians, and Han Chinese, among others. With a few historically interesting exceptions, each of these groups resembles its neighbors more than it resembles groups farther away. Though two groups may be sharply divided culturally, the mixing of their DNA inevitably blurs the genetic boundary between them.

Furthermore, every human group, when viewed on a long enough time scale, is a complex mixture of previous groups. Of course, every group likes to think of itself as historically ancient and unlikely ever to disappear, but history demonstrates otherwise. The English, for example, are a combination of the Beaker Folk of the Bronze Age; the Indo-European Celts, who arrived in Britain during the Iron Age; the Angle, Saxon, and Jute invaders of the first millennium; the Viking and Norman invaders of about 1,000 years ago; and the many peoples who have immigrated to England more recently. Even the Bushmen are undoubtedly the result of a complex and protracted mixing of groups, and today they continue to have a complicated intergroup structure of related bands and tribes.

Sometimes geneticists draw the relationships among human groups in the form of a tree, with Asians and Europeans branching off from Africans, American Indians branching off from Asians, and so on. But such trees are fundamentally misleading because they do not show the links among groups. Human groups are more like clouds forming, merging, and dissipating on a hot summer day.

This genetic mixing is one reason why physical features are not rigorously sorted among groups. No matter where you go in the world, once you become accustomed to the general appearance of the people, you notice again the tremendous diversity of human features. In any group, people have wider and thinner noses, lighter and darker skin, distinctive and generically shaped eyes. We tend not to see beyond the facial features that we use to identify "race," which is why we often think that the individuals belonging to a different group all look similar. But humans are as physically diverse in Botswana as in Norway or India. In fact, without the two obvious guides of skin color and eye shape, dividing people into groups would be difficult, and even with those markers we often draw the wrong conclusions.

It was not inevitable that human history would work out as it did. Suppose an archaic species of human beings such as *Homo erectus* had become firmly established in the Americas during a time of lower sea level. As the sea rose, modern humans might have been unable or unwilling to leave the Old World. In that case, when Columbus came ashore in the West Indies, he would not have encountered humans who had separated from his own ancestors in the relatively recent past. He would have met people who were clearly different from modern humans — slope-browed, linguistically primitive people with a cranial capacity about two-thirds of ours. Only in such a situation would it be possible to say that human beings could be divided into distinct groups.

Just as each group of humans results from the mixing of previous groups, so any given person is the descendant of many ancestors. But for individuals, this mixing is much more intricate.

Usually, when we think of our ancestors, we think about the branches of our family trees that we can extend the furthest into the past. For example, many people can trace the genealogical path responsible for their surname further than other ancestral lines. But this emphasis on a few lineages disguises the bushiness of our ancestry.

One generation ago each of us has two ancestors — our mother and father. Two generations ago we have four grandparents; three generations ago, eight great-grandparents, and so on. With each generation, the number of our ancestors doubles.

In this book I'll use twenty years as the measure of one human generation, so ten generations is 200 years. That's not long in the broad sweep of history: 200 years ago Thomas Jefferson was president of the United States, Beethoven was writing his Second Symphony, the Manchu emperor Qianlong had recently died, and the German naturalist Gottfried Treviranus had just coined the word "biology." Yet just ten generations ago, each of us had 1,024 great-great-great-great-great-great-great-great-grandparents (and most of us got our surname from just one of those individuals).

The doubling of ancestors continues back through time. Twenty generations ago, each of us had more than a million ancestors (1,024 × 1,024, to be exact, since each of our ancestors ten generations ago had 1,024 ancestors ten generations before that). Thirty generations ago — at the beginning of the fifteenth century — each of us had more than a billion (1,024 × 1,024 × 1,024).

This is getting a bit ridiculous. A billion people did not live on earth in the year 1400. (The actual estimate is around 375 million.) The exponential rise in the number of our ancestors must break down at some point.

Here's why. If you actually made a list of your 1,024 great-great-great-great-great-great-great-great-grandparents, most of the names on that list would be different. But some names would be the same, and for a simple reason — they would be the same person. These were the individuals who contributed to your ancestry through more than one line of descent — or through what I'll call circles of inheritance. Say one of your female ancestors several generations back had two daughters, each of whom married and had children. Now say that a child from each of those families married each other when they grew up (they would be cousins). The children of married cousins have only six great-grandparents, not eight, because one set of great-grandparents is the same on both sides. So when the children of married cousins make a list of all their great-grandparents, two names on that list appear twice.

This phenomenon becomes more frequent as one goes back through the generations. Marriages between first cousins are rare in many countries (although they remain common in others). But marriages between second cousins are more common (the children of such marriages have just fourteen great-great-grandparents rather than sixteen), and marriages between third cousins — if they even know they are third cousins — are more frequent still. The greater the number of generations taken into account, the greater the possibility that any two people who marry are distant cousins. By the tenth generation back, almost all of us have circles of inheritance in our ancestry.

Because we can't keep track of all our ancestors, most of us are unaware of how closely related we are to other people. But genealogical research can turn up surprising connections. For example, when a U.S. president is elected, professional genealogists begin researching his family, and invariably they turn up many circles of inheritance. Calvin Coolidge and his wife were eighth cousins twice removed. (In genealogical terms, "first cousins," "second cousins," and so on refer to people in the same generation, while "removed" refers to people in different generations.) Franklin Delano Roosevelt's parents were thirteenth cousins. Gerald Ford is the seventh cousin three times removed of Grover Cleveland; both are descended from the same seventeenth-century immigrant.

This kind of analysis is difficult to do for a particular person, because records of ancestors are hard to find (and often don't exist). But such an analysis can be done for a statistical "any person" by making certain assumptions and plugging numbers into a computerized model. The result is a tally of the number of an average person's different ancestors in each previous generation. The simulations show that within a few centuries the number of ancestors becomes extremely large, even if one allows for the effects of cousins marrying cousins. Forty generations, or 800 years, ago, each of us had many millions of ancestors.

At that point, the number of our ancestors reaches a critical threshold. It becomes greater than the total population in various parts of the world. For example, the populations of Europe and of Africa in the year 1200 were each about 50 million, and the population of Asia was about 250 million. In other words, our ancestors 800 years ago probably included much of the adult population in those regions of the world where they lived.

Joseph Chang, a statistician at Yale University, has recently shown that all of the people living in various regions of the world more than about 800 years ago can be divided into two categories. Each individual was either the direct ancestor of everyone in that part of the world alive today (about 80 percent of people fall into this category), or the lineage represented by a person went extinct, making that person an ancestor of no one today. This calculation requires random mating in the region of the world being analyzed — that is, each male has to have an equal chance of marrying any female in the region — and no one knows the extent to which this condition applies in the real world. But observations of typical family trees, plus the lessons of history, indicate that mating is probably close to random over long time periods.

Now factor in the consequences of human migrations. Say that 800 years ago one of our millions of ancestors was a recent immigrant to that part of the world. For Asians, that ancestor might have been a European adventurer who settled in a town on the Silk Road and married an Asian woman. So long as the children of that pair married other Asians, that European adventurer would today be an ancestor of the entire population of Asia. Or suppose that an emissary from Ethiopia married a woman in the court of Henry II and had children. Today, all Europeans are descended from that Ethiopian — and from other Africans who married into the European population and had children in medieval times. Similarly, all Africans today descend from Chinese traders who visited Africa, and undoubtedly fathered children, early in the 1400s.

Genealogically speaking, when a person marries into a family, he or she marries the whole family. In other words, if a member of a particular human group was one of our ancestors, then all of that person's ancestors are also our ancestors. So if 800 years ago our ancestors included even a single European, African, or Asian, then 1,600 years ago our ancestors included most of the adult population of all three continents.

People like to trace their ancestry to famous figures. In Japan many families trace their lineages to the ninth-century emperor Seiwa. Many French believe they are descended from Charlemagne. Such claims are typically impossible to prove, and many genealogists dismiss them as fantasies.

However, these claims are almost certainly true. The exponential growth in the number of our ancestors going back in time connects us tightly to the past. If a historical figure who lived more than 1,600 years ago had children who themselves had children, that person is almost certainly among our ancestors. Everyone in the world today is most likely descended from Nefertiti (through the six daughters she had with Akhenaton), from Confucius (through the son and daughter he is said to have had), and from Julius Caesar (through his illegitimate children, not through Julia, who died in childbirth). One need go back only a couple of millennia to connect everyone alive today to a common pool of ancestors.

Being descended from someone doesn't necessarily mean that you have any DNA from that person. Essentially, the inheritance of DNA is weighted by numbers. If 800 years ago most of a person's ancestors were African and a few European, then most of that person's DNA comes from the African ancestors alive at that time. The amount of DNA each of us gets from any one of our 1,024 ancestors ten generations back is minuscule — and we might not get any DNA from that person, given the way the chromosomes rearrange themselves every generation. Still, the basic point is unchanged. The DNA now in our cells consists of bits and pieces of the DNA that was in thousands of people's cells a millennium or two ago. Our DNA is a patchwork quilt stitched together from the DNA of our ancestors.

The extravagance of our genetic ancestry needs to be kept in mind when thinking about the origins of our mitochondrial DNA. Sure, most Europeans received their mitochondrial DNA from a handful of women who lived in the past (one geneticist calls these mitochondrial ancestors the Seven Daughters of Eve). But today's Europeans received the DNA in their chromosomes, as opposed to their mitochondria, from thousands of other prehistoric Europeans, Africans, and Asians who lived at the same times as their mitochondrial ancestors. Similarly, the mitochondria of everyone on the planet today came from mitochondrial Eve, but the DNA in our chromosomes came from thousands of people who lived at the same time as mitochondrial Eve. The roots of our family trees extend to many of these early modern humans, not just one.

Some people might like to believe that the genetic mixing of people from different groups is rare — and that their ancestors certainly didn't mix with the hoi polloi. But groups have many ways of mixing.

An example is what geneticists call nonpaternity. Everyone has a biological father and a biological mother. But are our biological parents who we think they are? We can be fairly confident about our mothers, grandmothers, and so on back through the maternal line. Anyone who has watched how quickly an identification bracelet is snapped onto the wrist of a newborn baby knows that mothers and their children are rarely separated. Of course, mistakes can happen, as occasional newspaper articles and court cases attest. Two babies might lose their bracelets at the same time and have them switched. A well-meaning nurse might swap babies out of some vague sense of cosmic justice. But the percentages of mistaken maternity must be low.

The percentages of mistaken paternity are not nearly so low. Medical students are taught that 5 to 10 percent of the fathers identified on birth certificates are not the true biological fathers. Geneticists who do large-scale surveys that incidentally reveal paternity confirm this number, though few have been willing to publish their results. The genetic studies that have been published find nonpaternity rates ranging from a few percent to well into double digits. One intriguing, if not entirely unexpected, finding: nonpaternity tends to be higher for first-born and last-born children.

Given the huge number of ancestors we all have, questions about the male contributions to our ancestry are a consideration for all families. I have a good friend from an extended American family that descends on the male side from Swedish ancestors. But the family name is an unusual one, so one of my friend's cousins decided to do some research. He traveled to Sweden and, after making a large number of inquiries, discovered that the family name came from a large farm outside Stockholm where prostitutes used to send their children to be raised.

Nonpaternity is one way in which ancestries become mixed, but there are many other ways. Adoptions bring new sets of genes into families and groups. Multiple marriages further scramble genealogies. The constant flux of people and groups ensures that human gene pools will always be mixed. Though populations of modern humans are not panmictic, they are not and never have been genetically isolated.

Given our intricate genetic histories, what does it means to belong to a human group?

First of all, most of the groups to which we belong have nothing to do with biology. We are bowlers, farmers, veterans, cab drivers, politicians, dancers, soldiers, divorcees, beggars, nuns, computer technicians, retirees. Perhaps a group originated among people of a certain ancestry, but unless it enforces membership restrictions, the association between the group and ancestry tends to blur over time. Islam, for example, originated among Arabic people, but today the majority of Muslims are Asians, not Arabs. Most human groups are the products of culture, not biology.

"Racial" and ethnic groups are also products of culture. When we label someone an Egyptian, Eskimo, or Asian, we are using physical and cultural characteristics to put that person in a culturally defined category. Without such categories, we would have to describe the person as having medium-dark skin or an epicanthic fold or an extra layer of epithelial fat, which really tells us nothing of any interest about that person.

Some people use this observation to conclude that all human groupings are cultural inventions. That's obviously absurd. Real physical differences exist between the average Nigerian, the average Norwegian, and the average Filipino. Most of the members of these groups share a common biological history, which is reflected in their DNA.

But modern humans are too young as a species and have interbred too enthusiastically to develop substantial genetic differences. For example, geneticists have never found a mutation that is 100 percent present in any "race" or ethnic group and 100 percent absent in another, and given our genetic history they never will. In terms of our DNA, all humans overlap.

This perspective on human groups provides a valuable tool with which to examine the genetic connections between the English father and South African mother of Sibongile Makhaya, the young woman introduced at the beginning of this chapter.

At some point in the past, a woman who probably lived in eastern Africa had two daughters. Some of the female descendants of one of these daughters eventually migrated into the Middle East and then into Europe. This line of female descent carried on for many millennia in an unbroken chain into the twentieth century, when a woman in this line gave birth to Sibongile's father.

The female descendants of the other daughter remained in Africa. Even when this lineage was young, groups of modern humans within Africa had probably begun to diverge. According to research done by geneticists and anthropologists at the University of Utah, modern humans appear to have separated into several subgroups before some of them left Africa. Because of their separation, these groups developed some amount of genetic differences before the migrations from Africa began. The Utah researchers call this hypothesis the "weak Garden of Eden model."

It is tempting to equate these early subdivisions within Africa with the three major groups of Africans who live south of the Sahara today — Bushmen, Pygmies, and the rest of the population. ("Pygmies" is another disparaging term for which no good replacement is available; like the Ju/'hoansi, subgroups of Pygmies have names for themselves, but they have no word for the larger group. In parallel with the term "Bushman," I'll refer to the Pygmies, somewhat inaccurately, as Forest Dwellers.) Not enough is yet known, however, to figure out the genetic relationships among these groups. The very early history of modern humans in Africa remains largely a blank, in part because human fossils tend to disintegrate in the acidic soils of tropical forests. And even when fossils between 15,000 and 150,000 years old have been found, linking the bones with existing groups has proved difficult.

Still, the genetic evidence is suggestive. Judging from their mitochondrial and Y haplotypes, the ancestors of the Bushmen had to be one of the first groups to become established in Africa. Similarly, several tribes of Forest Dwellers, who today live in scattered remnants in central Africa, have very old mitochondrial and Y lineages. And like the Bushmen, the Forest Dwellers seem to have been much more widely distributed at some time in the past. Documents from ancient Egypt mention them living south of the country along the White Nile, far from where they live today.

Africans described by the term "black" tend to have more recent mitochondrial and Y haplotypes, though it is unknown where and when these haplotypes diverged from the ancestral types. I prefer not to use biological terms to label human groups, so I'll refer to these people as Central Africans, though today they are scattered throughout the continent. Central Africans are by far the largest and most widespread group in Africa, but in earlier times their distribution was more limited. They seem to have lived predominantly in the fields and savannas north of Africa's equatorial rain forests.

As is always the case with modern humans, these various African groups were never entirely distinct. For example, Central Africans and Forest Dwellers have clearly mixed in various places, producing "Pygmoid" populations with characteristics of both groups. Also, all three groups have expanded and contracted over time. One of the largest such movements — indeed, one of the most dramatic population movements of all time — began about 2,500 years ago. Somewhere around the border between present-day Nigeria and Cameroon, just where the west coast of Africa swings southward, a group of Central Africans began to expand and move into areas occupied by other peoples. These newcomers spoke languages that belong to what is called the Bantu family, all of which use the word *bantu* to mean "people." Technically, Bantu is a linguistic rather than a demographic term. But this historical event is usually called the Bantu expansion, and the people are grouped under the category Bantu. (As with other such designations, this word has some negative connotations — especially in South Africa, where the former government used it to refer to all of the non-European population.)

The original causes of the Bantu expansion remain obscure, but two innovations greatly strengthened it. The first was the domestication of crops such as yams that are suited to the hot, wet summers of central Africa. The second was the introduction — or perhaps independent development — of iron smelting in central Africa. These two developments strongly reinforced each other. The smiths of central Africa produced the hoes, picks, and axes that Bantu speakers used to

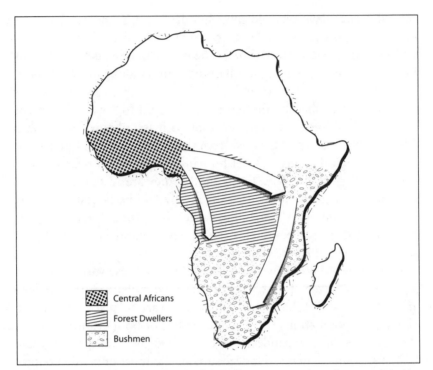

About 3,000 years ago, Africa south of the Sahara seems to have been occupied by three main groups: Central Africans, Forest Dwellers, and Bushmen. Then a sub-group of Central Africans began expanding east across the center of the continent and south along both coasts, pushing the Forest Dwellers and Bushmen into less productive areas.

clear forests and work their fields. In turn, the felled trees fed the fires of the smelters. With these twin innovations, the Bantu speakers created a technological juggernaut. They overwhelmed the Forest Dwellers and Bushmen, forcing these foraging populations to retreat into rain forests and deserts unsuited to agriculture.

Initially, the Bantu speakers spread east across the top of the Congo basin to the headwaters of the Nile around Lake Victoria. There they turned south, expanding along the east coast of the continent. Meanwhile, other Bantu-speaking groups spread southward down the west coast of Africa. Various groups crossed east to west or west to east across the continent, producing complicated waves of population movement. But Bantu speakers tended not to settle in the drier interior regions or along the coast of southern Africa, where their moisture-loving summer crops did not flourish. These are the areas where

Bushmen and the descendants of European immigrants are concentrated today.

The DNA of modern Africans clearly reflects this Bantu expansion. Himla Soodyall and her mentor, Trefor Jenkins of the University of the Witwatersrand, have tracked one particularly informative marker: a missing piece of mitochondrial DNA nine base pairs long. The deletion occurs in a portion of the mitochondrial DNA that doesn't do anything, so it has no effect on a person's body. But this mutation has spread across the continent in a clear swath. The deletion is rare among the people living in the original homelands of the Bantu speakers. It also is uncommon among the Bantu speakers who moved down the west coast toward Namibia. But Forest Dwellers from the Congo and the Central African Republic have it, and it is especially frequent among Bantu speakers who moved down the east coast of Africa.

As with all mutations, this nine-base-pair deletion originally must have occurred in a single person. Given its current distribution, it appears to have arisen either in a Forest Dweller woman whose descendants then mixed with the passing Bantus or in a Bantu speaker, some of whose female descendants mixed with and remained with the Forest Dwellers.

Here we can rejoin the mitochondrial lineages of Sibongile Makhaya's ancestors. As the Bantu speakers expanded across Africa, they divided into chiefdoms, including the Sotho and the Zulu. Sibongile's grandmother on her mother's side was Sotho. Her mother and grandmother — as well as Sibongile herself — therefore may have inherited from their Bantu-speaking ancestors the mitochondrial DNA with the nine-base-pair deletion.

This mitochondrial DNA had passed through many previous populations. It existed among the Central Africans who produced the Bantu speakers. They inherited it from the modern humans who moved west to that part of the continent from eastern Africa. And ultimately it came from the African woman who is at the root of the separate mitochondrial lineages that led into both Europe and western Africa.

Sibongile's English father and South African mother were therefore distant cousins through their common mitochondrial ancestor. When they produced Sibongile, they closed a circle of inheritance that had opened many millennia before.

3 ➤➤

The African Diaspora and the Genetic Unity of Modern Humans

We hold these truths to be self-evident, that all men are created equal.

— Thomas Jefferson, *Declaration of Independence of the United States of America*

THE CITY WHERE I LIVE, Washington, D.C., is one of the most diverse in the world. More Ethiopian expatriates live in Washington than in any other city. The population of Vietnamese, Cambodians, and Laotians exceeds 50,000. One of every eight Washington residents was born outside the United States. Within one downtown block, you can order food from the national cuisines of Ghana, India, Thailand, Malaysia, China, Japan, Italy, France, El Salvador, and Brazil.

But by far the largest ethnic group in Washington consists of African Americans. Of the 570,000 people who live in the District of Columbia proper, 60 percent are African American, as are about 25 percent of the metropolitan area's 5.5 million people. Many are the descendants of African Americans who moved to Washington from the southern United States after World War II in search of better jobs and better schools for their children. Some can trace their ancestry back to African Americans, free or slave, who lived in and around Washington before the Civil War.

The suffix "American" in this chapter has a particular connotation. The previous chapter used the term "Central African" to refer to the largest population group in Africa. Many of the ancestors of today's

African Americans were indeed Central Africans. But almost all African Americans also have a large number of European-American ancestors. On average, one-fifth to one-quarter of their DNA comes from Europeans. If any African American other than a recent immigrant were to construct a family tree for the past four hundred years, a substantial fraction of the names on that tree between ten and twenty generations ago would be individuals of European ancestry.

By the same token, many European Americans have relatively recent African-American ancestors. In some cases the ancestor was one member of an openly interracial marriage. But in most cases the ancestor was an African American whose skin was so light that he or she was passing as a European American. These "passers," as they are called, have been a powerful force for demographic mixing in the United States. Several of the children of Thomas Jefferson and his African-American slave Sally Hemings eventually blended into the majority population. Certain cities in the southern United States had large numbers of passers, whose descendants gradually spread into the rest of the country. Some light-skinned African Americans entered the armed forces, were classified as whites, and joined all-white units. Once discharged, many simply continued to define themselves as white.

Thus the suffix "American" here implies an especially high degree of genetic mixing. Such mixing is common in many parts of the world: South Africa is a good example of how people from different continents can produce highly varied populations. But the United States' reputation as a melting pot is borne out by the DNA of its citizens. Few Americans today can claim that all of their ancestors several generations back were from a particular part of the world.

Given this extensive history of mixing, the strength of racial prejudice in the United States can seem perplexing. Throughout the country's history, Americans have drawn rigid distinctions between black and white, Indian and European, Asian and non-Asian, Latino and Anglo. Furthermore, these distinctions have been rooted in the belief that sharp genetic differences separate groups, differences that shape behavior as well as appearance.

Scientific research has shown that these claims have no merit. Given the history of our species, the behaviors characteristic of a group must be the product of culture — of what people learn — not of genetics.

That these behaviors seem so ingrained is not a measure of our biological heritage; it is a measure of history's power to shape the collective consciousness of nations and peoples.

All non-Africans descend from Africans who left the continent within the past 100,000 years. But the flow of people out of Africa must have waxed and waned over time. Once the Middle East was settled by modern humans, the people there would have presented an obstacle to new migrations, because incoming groups would have had to compete with them for resources. The prehistoric peoples of the Middle East and the Nile River delta undoubtedly exchanged mates. But the geography of the Old World — especially the barrier posed by the Sahara Desert and the fact that everyone traveling from Africa to Eurasia has to pass through the narrow isthmus of the Sinai — suggests that in prehistoric times people moved between the continents in rather low numbers.

More people journeyed between Africa and Eurasia following the rise of civilizations in the Middle East. Black Africans from south of the Sahara began serving as merchants, sailors, servants, and soldiers throughout the Mediterranean world. The invention of oceangoing ships gave people a new way to move between continents. Many Africans settled in Europe and Asia and married non-Africans, and their descendants gradually were absorbed into the population. Even today, mitochondrial and Y haplotypes characteristic of Africa are found in people throughout Europe and Asia, and some of the haplotypes date to these early African exoduses.

The early movements of people out of Africa were dwarfed by the population movements associated with the international slave trade. Slavery was a common feature of ancient societies; in classical Greece and Rome, slaves constituted as much as 30 percent of the population. But slave masters in the ancient world did not discriminate by geographic origin. In Rome, African slaves worked alongside slaves from Iberia, Gaul, northern Europe, Thrace, Sarmatia, India, the Arabian peninsula, Egypt, and Carthage. In fact, many of the early slaves were destitute Romans who had sold themselves into slavery to pay off debts.

In the Middle Ages the geographic origins of slaves in the western world began to change. The number of slaves from Europe and Asia gradually fell, while the number from Africa grew. By 1700, Africa was the world's primary source of slaves.

Scholars have cited many reasons for Africa's rise as a slave producer. Tropical climates and poor soils in Africa limited agricultural productivity, so the overall economic return from selling a young male into slavery was greater than if that man became a farmer. Africans captured and sold slaves in exchange for manufactured goods, which delayed the continent's economic development and further depressed the value of labor within Africa. The rise of sugar production in the New World created a need for huge numbers of agricultural laborers, and Africans were more resistant to the diseases of tropical sugar plantations than were people from other parts of the world. The slave trade was highly profitable to those who provided, transported, and used slaves, and the continent's political fragmentation made it difficult for African leaders to stop the practice. All of these factors tended to reinforce one another. Once Africa became established as the leading exporter of slaves, the trade gained a momentum that was very difficult to reverse.

The slave trade was responsible for one of the largest human migrations the world has ever seen. Even before Europeans began shipping African slaves to the New World, millions were sent to Europe, the Middle East, and as far away as China. Sizable communities of sub-Saharan Africans arose in cities from Lisbon in the west to Hyderabad in the east, from London in the north to Cairo in the south. Some of these communities have survived to the present, but in most places Africans gradually intermarried with non-Africans and blended into the surrounding population.

The flow of Africans to the New World eventually exceeded that to the Old. Between the early 1500s, when the first slaves were transported directly from Africa to the Americas, and 1870, when the last verified shipment of African slaves made landfall in Cuba, approximately 12 million enslaved Africans traveled across the Atlantic. Africans quickly became a major portion of the population in the Americas, especially as indigenous populations were decimated by Old World diseases. As late as 1800, several times as many Africans as Europeans lived in the New World.

Perhaps 2 million Africans died during the hellish "Middle Passage" from Africa to the Americas (so called because it was the middle sec-

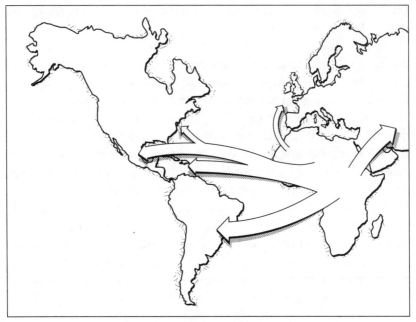

Approximately 10 million African slaves traveled to the New World between the early 1500s and 1870, with the largest numbers (as indicated by the size of the arrows) going to Spanish America, the Caribbean, and Brazil. An estimated 5 million enslaved Africans were sent to the Old World.

tion of a triangular trade route from Europe to Africa, Africa to the Americas, and the Americas back to Europe). Chained together in tiny, fetid spaces under unimaginably horrible conditions, they died of dysentery, smallpox, hunger, thirst, suicide, and suicidal rebellions. A slave who exhibited the first signs of sickness was often thrown overboard to prevent an epidemic. Others willingly jumped overboard to drown and escape their captors.

By far the majority of Africans shipped to the New World disembarked in South America and the Caribbean. (The sizes of the arrows on the accompanying map give a rough indication of the number of people transported to different areas.) The mortality of slaves shipped to these areas was shockingly high. Working a slave to death and buying a new one was cheaper than supporting the development of long-term communities and families. But many slaves did survive, and many eventually gained their freedom. By the time slavery was abolished in Brazil in 1888, nearly half the population of the country had ancestors who had been slaves.

The British colonies of North America received far fewer slaves than did Central and South America. But mortality was lower than in South America and the Caribbean (although still much higher than it was for Europeans), and Africans were always a major fraction of the nonnative population. At the time of the first official U.S. census in 1790, about 20 percent of the country's 4 million people were African American. (As of the 2000 census, African Americans represented 13 percent of the total population of 280 million.)

The large-scale emigration of Africans, Europeans, and, later in the nineteenth century, Asians to the Americas created a human medley unprecedented in history. In the Old World, skin color, facial features, and other physical characteristics have tended to vary continuously. As a result, people in Europe, Africa, and Asia generally interacted with people who were much like themselves physically. Only in the mixing bowl around the Mediterranean and on long-range trade routes did groups that looked substantially different from each other frequently come into contact.

In the New World, the situation was the reverse. The three major groups of American immigrants came from geographic extremes of their respective continents — western Africa, northwestern Europe, and eastern Asia. These groups were overlaid on a fourth group — Native Americans — that had Asian roots but differed substantially from Chinese and Japanese populations. A mischievous god moving groups of people around the globe would have a difficult time assembling a more disparate collection.

The physical differences among the new occupants of the Americas did little to slow the natural human tendency to interbreed. Today most of the people of South and Central America are of mixed European, American Indian, and African descent. In Brazil, for example, African slaves had more ways of becoming free than in other countries, and the settlers from Portugal had a history of absorbing people from elsewhere. As a result, intermarriage has been extensive, and the physical characteristics of Brazilians today extend across a broad range. Racism is not absent in Brazil, despite claims that all races are blending. People with darker skin have more trouble finding good jobs, good housing, and good schools. But skin color in Brazil varies along a continuum, and social distinctions vary continuously as well.

Things turned out differently in the United States, where slave owners and their political allies had a strong interest in maintaining a sharp distinction between African Americans and European Americans. Children born of slaves and slave owners represented a blurring of the distinction between slave and free, between property and property owner. To deal with this problem, these children almost always were treated as black. Thus arose the "one drop" rule, which held that having a single black ancestor made a person black. Today, a strict application of this rule would make a substantial fraction of all Americans African Americans. But it served the purposes of slave owners eager to add to their holdings.

Laws reinforced this distinction. As early as 1691 a statute in colonial Virginia barred "[male] negroes, mulattos and Indians intermarrying with English or other white women," conveniently overlooking the fact that relations between European-American males and African-American females were much more common than those prohibited by the statute. Later laws in other states, including many in the North and the West, banned all interracial marriages, including those between people of Asian and European descent. Not until 1967 did the U.S. Supreme Court declare such bans unconstitutional.

These laws reflected a powerful sense of separateness in the United States. Despite the continuous mixing of Africans and Europeans, many European Americans displayed a strong psychological need to believe that the two groups were fundamentally different. This belief accorded with Europeans' view of their place in the world, which arrayed various non-European peoples beneath the exemplar of European civilization. It also gave Europeans a way to justify their barbarous treatment of African slaves.

Many scientists lent their support to this cause. Throughout the Enlightenment, scientists studied different human groups in an effort to put them into discrete categories. In 1758, for example, the Swedish botanist Carolus Linnaeus gave the human species its formal name, *Homo sapiens*. He then divided the species into four subcategories: red Americans, yellow Asians, black Africans, and white Europeans. He described *Homo sapiens americanus* as "ill-tempered, . . . obstinate, contented, free." *Homo sapiens asiaticus* were "severe, haughty, desirous." *Homo sapiens afer* were "crafty, slow, foolish." And *Homo sapiens europaeus* were — of course — "active, very smart, inventive."

Over the last few decades, scientific efforts to divide human beings into discrete categories have all but collapsed. The categories were clearly artificial, since all groups blend into each other. And the campaign to define human races came to be seen for what it was: a misguided attempt to use the methods of science to excuse the inexcusable.

The failure to define races scientifically has not ended racism. On the contrary, the pronouncements of racist organizations and individuals are as bigoted as ever. And a pervasive if less blatant racism remains deeply embedded in the broader society, despite all the progress that has been made against prejudice. To take just one example, racist groups have never been particularly large in absolute terms. But their activities receive great media attention because they play on fears that cut deeply into the American psyche.

These fears also have made the United States susceptible to another kind of media phenomenon. Every few decades the country undergoes a spasm of self-doubt occasioned by the publication of a book purporting to reveal innate mental differences among groups. These books make more or less the same arguments. They contend that particular groups in U.S. society have inborn traits that make them inherently less able to succeed. Inequities between groups therefore do not reflect differing access to education, good jobs, and decent housing. They reflect genetic differences that are impossible to change.

Even before much was known about the genetics of human groups, these arguments were unconvincing. Take the most notorious example — the differences in IQ scores among African Americans, European Americans, and Asian Americans. The average scores of African Americans and European Americans in the United States typically differ by ten to fifteen points, while Asian Americans tend to score about ten points above European Americans. Hereditarians repeatedly cite these scores as proof that groups have different genetic capabilities. Yet the available evidence points instead toward the overwhelming influence of environmental factors:

- When IQ tests were developed in the early twentieth century, people from different parts of Europe had different average scores. At that point, hereditarians used these differences to distinguish among Nordics (northern Europeans, roughly speaking), Alpines (eastern and central Europeans), and Mediterraneans (southern Europeans). Today the descendants of these immigrants score equally well on IQ tests. Yet the same arguments are now applied, through a sort of bracket creep, to African, European, and Asian Americans.

- If genetics were the cause of IQ differences, then African Americans who have higher proportions of European genes should score higher on IQ tests than those with fewer European ancestors, but no such effect has been found. The critical difference is not whether a person is genetically African American; it is whether a person has been treated as an African American.

- Throughout the twentieth century, IQ scores have been going up for all groups, according to a variety of tests conducted in many different countries. These rises have occurred much too rapidly to be the product of genetic changes. They must result from better diets, better health care, and better education.

- Many studies show that children who receive good prenatal care and early childhood education on average score higher on IQ tests than children who do not. Since proportionally more African Americans than European Americans live in poverty in the United States, their scores on IQ tests tend to be lower.

- When researchers tracked down the children born to German mothers and U.S. soldiers during the Allied occupation of Germany in World War II, they found no difference in the IQ scores of children with African-American versus European-American fathers.

- Minorities in many countries score lower on IQ tests than do the majorities, regardless of their ancestry. An example is the Buraku of Japan, a minority that is severely discriminated against in housing, education, and employment. Their children typically score ten to fifteen points below other Japanese children on IQ tests. Yet when the Buraku immigrate to other countries, the IQ gap between them and other Japanese gradually vanishes.

Figuring out why some groups score lower on IQ tests than others is not difficult. Disadvantaged social groups must shoulder many burdens, some of which are imposed on them, some of which they create for themselves. They may have higher dropout rates, lower incomes, more illegitimate births, higher levels of substance abuse, fewer opportunities in housing and employment, and so on. With such liabilities, it makes little sense to pin group differences on biology. But that has always been the easy way out. If group differences are believed to be inborn, efforts to improve the lives of the disadvantaged are unnecessary because they inevitably will fail.

That the supposed influence of genes on intelligence evaporates under closer scrutiny is not surprising. People are too genetically similar to have developed the kinds of intelligence differences cited by hereditarians.

By comparing the DNA sequences of people from many locations around the world, geneticists have been able to measure the genetic differences between individuals and between groups. What they have found is that about 85 percent of the total amount of genetic variation in humans occurs within groups and only 15 percent between groups. In other words, most genetic variants occur in all human populations. Geneticists have to look hard to find variants concentrated in specific groups.

The pattern is quite different in other large mammals. Among the elephants of eastern and southern Africa, 40 percent of the total genetic differences occurs between groups. For the gray wolves of North America, group differences account for 75 percent of the total genetic variation. Most conservation biologists hold that group genetic differences have to exceed 25 to 30 percent for a single species to be divided into subspecies or races. By this measure, human races do not exist.

Still, human groups do have different frequencies of some genetic variants — that's how geneticists can track human history using our genes. These differences provide people committed to a racial world view with one last argument. If groups can differ in appearance for genetic reasons, they say, why can't they differ as well in temperament, character, or intelligence? For example, maybe cold climates exerted some sort of selective pressure on the brain as well as on skin color. If people had to be smarter to survive in far northern areas, maybe intelligence was somehow concentrated in their descendants.

The argument doesn't work, for two reasons. First, no mechanism has been identified that could sort complex attributes within such a genetically homogeneous and interconnected species. The idea that natural selection favored different cognitive traits on different continents seems designed more to justify social prejudice than to establish testable hypotheses. The people most adapted to the rigors of the European climate were the Neandertals, not modern humans. Population levels declined several times in Africa following the appearance of modern humans, and the declines would have imposed the same sorts of selective pressures on Africans as those that non-Africans presumably faced. Animal species with worldwide distributions aren't more intelligent in cold climates than in warm ones. No evidence at all exists that different human groups have ever been under different selective pressures for cognitive traits.

Even if a potential differentiating mechanism could be identified, the case for group differences fails for a second reason. A fundamental distinction exists between a simple trait such as skin color and a complex cognitive attribute such as intelligence. Skin color is determined by a handful of genes and does not depend on the experiences a child has in the womb and while being raised. The development of the brain involves thousands of genes and is indissolubly linked to experience. In a species as genetically homogeneous as ours, showing that the distinctive behaviors of a group have a biological origin will never be possible, because the complexity of the gene-environment interaction will always dwarf any genetic difference between groups.

This point needs to be emphasized. Because of the intricate links between genes and experience, no one can say with certainty that a particular cognitive trait is, say, 50 percent environmental and 50 percent genetic. The two are so intertwined that they cannot be separated. Even if a particular trait were found to be genetically influenced in a particular environment, that would say nothing about the genetic influence on that trait in other environments. And because we cannot test genetically different people in the full range of human environments — we can't even imagine the full range of human environments — we never will be able to say that a particular genetic variant always causes a particular complex behavior.

This will be a contentious issue over the next few decades. As more is learned about how genetic variants differ in frequency among groups, people will try to link those differing frequencies to the cultural attributes of groups. The reality will always be much more complicated than these simple correlations would indicate. But these efforts will receive great attention because they accord with popular prejudices.

The desire to attribute complex behaviors to unknown genetic forces is somewhat puzzling. After all, individuals and groups differ for many obvious reasons. Individuals have unique experiences from the moment of conception. They receive varying levels of nutrition and medical care. They are not treated uniformly by adults, teachers, and peers. They are born into cultures with particular histories and beliefs. Why ascribe group differences to genetic forces when the agents of differentiation are right in front of our faces?

One last issue must be addressed before we can dismiss for good the notion that groups of people differ biologically in fundamental ways. Physicians have known for a long time that some groups are more prone than others to particular diseases. African Americans tend to have a higher prevalence of high blood pressure, asthma, and prostate cancer than European Americans. Hispanics in the United States are more likely to suffer from diabetes. Cystic fibrosis occurs predominantly in whites of European descent; Tay-Sachs disease, in Jews of Eastern European descent.

Almost all of the disease susceptibilities in human groups clearly result from environmental influences. Japanese men living in Japan rarely get prostate cancer, but their rates rise if they move to the United States. Though many African Americans have high blood pressure, the people of rural West Africa have the *lowest* blood pressure of any group in the world. Most of the health risks we encounter come from our environment, not from our genes. We smoke, drink alcohol, eat unhealthy foods, don't exercise enough, are exposed to carcinogens at home or in the workplace — and we're all susceptible to simple bad luck.

But genetic differences among groups do have a small influence on health. This is most obvious with diseases caused by mutations in single genes, such as sickle cell disease, cystic fibrosis, or Tay-Sachs, but group genetic differences also contribute to more common diseases, including cancer and heart disease. Could these genetic influences on health point toward other kinds of group genetic differences, including behavioral differences?

Georgia Dunston, the chair of the Department of Microbiology at Howard University in Washington, D.C., has been thinking about these issues for a long time. "I've been interested in human variation since I was a girl growing up in the segregated South," she says. "I've had some questions that I've wanted to ask God about the differences between people, given that he loves everyone equally."

Dunston was born and raised in southern Virginia in a working-class African-American family. After graduating from Norfolk State University with a degree in biology, she moved to New York City and tried to get work as a lab technician. When no lab jobs turned up, a college adviser recommended her at the last minute for a master's program at Tuskegee University. In her final year at Tuskegee, though Dunston had never considered the possibility of pursuing a doctoral degree, an exchange professor from the University of Michigan talked her into applying for that university's Ph.D. program. She was the first African-American doctoral student in human genetics at Michigan and is one of just a handful of African-American professors doing human genetics research in the United States today.

She chose Howard, a historically black university, because she wanted to work on problems important to the African-American community. She immediately dove into one of the most contentious: the science and politics of organ transplantation. For reasons that are still unclear, African Americans with high blood pressure or diabetes suffer from kidney failure more often than do European Americans. African Americans therefore constitute a disproportionate number of the people on waiting lists for transplanted kidneys.

The problem is that human kidneys are not totally interchangeable. Particular kinds of molecules on the outsides of all human cells differ among people much the way blood types do, and they serve a somewhat similar function. If the tissue types of an organ donor and a recipient differ too much, a transplanted organ is much more likely to be rejected.

When Dunston was setting up her research laboratory at Howard in the mid-1970s, organ rejection was a manageable issue for European Americans. Almost all had easily identifiable tissue types, so it was not difficult to match organs with patients. But when geneticists tried to type the cells of African Americans, they often encountered "blanks" in the data, because the available tests were unable to identify many African-American tissue types. As a result, more European Americans were receiving the available kidneys, and when African Americans did receive kidneys, their rejection rate was higher.

With her colleagues at Howard and another team of geneticists at Georgetown University, Dunston helped broaden the range of detectable tissue types. Within a few years, the number of blanks in tissue-typing tests declined significantly, and the available organs and patients could be matched up more precisely. But working on this problem made Dunston think about why African Americans have a broader range of tissue types than European Americans. Why had human history produced this effect?

When the mitochondrial DNA data identified Africa as the homeland of modern humans, the picture suddenly became much clearer. "Most of human evolutionary history occurred when we were all Africans," says Dunston. "If variation represents changes in genes over time, then the finding of greater variation in African people is consistent with Africa being the birthplace of humankind, and with African groups as the oldest human populations. Using European Americans as the standard for assessing human genetic variation is like using the tip of a branch to measure the size of the trunk. Europeans are one of the youngest populations, so of course they're going to be more homogeneous than African groups. But the greater heterogeneity of Africans is not an abnormality except when you're trying to explain variation in the population on the basis of the narrower spectrum of variation in Europeans."

In short, Dunston realized that the genetic history of a group can be a key consideration in understanding the origins of disease. Soon she began to apply this insight to other problems. For example, in work done at the National Institutes of Health, Dunston and a group of other researchers have been looking at a specific mutation that occurs in a gene called BRCA1. Women who have harmful mutations in BRCA1 are more likely to develop breast or ovarian cancer at a young age, although the effect is complicated: most women who develop breast cancer do not have a mutated BRCA1 gene, and not all women with the mutated gene suffer from breast cancer. Still, mutations in BRCA1 are associated with cancer through a mechanism that remains unknown.

Dunston and her colleagues have been looking at a group of women with a particular BRCA1 mutation — an insertion of ten nucleotides that destroys the gene's function. The group includes African women from the Ivory Coast and African-American women from the Bahamas, the southeastern United States, and Washington, D.C. Given the current distribution of the mutation, Dunston and her coworkers believe that it occurred in someone who lived in western Africa 200 to 300 years ago. Some of that individual's descendants stayed in Africa, while others came to the Americas through the slave trade.

Any time a mutation occurs in DNA, it creates a new haplotype. By looking for that haplotype in current populations, geneticists can trace its spread. In the first few generations after a mutation occurs, it tends to stay within the group in which it originated, because mating occurs largely within the group. But as individuals from different groups begin to interbreed, the mutation spreads. That process has started with the BRCA1 mutation studied by Dunston and her colleagues, but the defective gene is still found largely among Africans and African Americans.

Meanwhile, similar mutations are occurring in other groups. For example, many other mutations have been found that impair the BRCA1 gene's function, and some are still confined largely to specific groups. Finns have one set of mutations, Japanese another, and Eastern European Jews a third. As with the mutation that arose in western Africa, new mutations tend to remain localized in the group. But over time, as groups mix, so do their distinctive mutations.

Dunston's work on BRCA1 exhibits several of the features that can be expected of continued research into the links between genes and disease. First, the effects of BRCA1 mutations on cancer susceptibility are relatively small. Any complex human trait is shaped by a wide variety of genes, as well as by experience and chance events. The effect of any one gene on that trait is therefore limited.

BRCA1 also shows that mutations occur throughout the human population, not just in specific groups. Though specific mutations necessarily occur within individuals and spread initially within groups, those mutations are not confined to a group for long. Because humans are always mixing, the human gene pool is always interconnected.

To Dunston, these new insights emerging from genetics research demand a wholesale revision of our ideas about race. "Here in the United States, we've equated racial groups with biological constructs and have used race as a surrogate for biology," she says. "But concepts of race based on appearance are too crude for biology at the level of the human genome. The human genome is not partitioned according to any definitions of racial groups. In fact, there are no parts of the genome that define every member of a racial or population group. The number of genes that have changed to produce all of the variation observed in different racial groups is minuscule compared with the backdrop of natural genetic variation in humans. We have to revisit our concepts and definitions of self in light of the new knowledge based on a more comprehensive view of the whole genome.

"Genetics research is about seeing the whole genome and the interdependence of its parts," Dunston says. "It's about health as a dynamic and harmonious balance of interacting genomic systems. It's about the order and complexity of life expressed at multiple levels of organization. And it's about human identity and our connections to all of humanity through the genome. It's a new knowledge base that we can use to answer the questions that we have within ourselves: Who am I? Why am I here?"

Using the genetic differences between individuals and groups to reconstruct human history, as I do in this book, can be a dangerous undertaking. It may seem to imply that the genetic differences between groups are as substantial as the cultural differences that divide us. It even may seem to imply that genetic factors are the cause of these cultural differences.

To avoid these misinterpretations of genetic information, we need to think carefully about what our genes are saying. First, we have to keep in mind the extreme fluidity of human groups. The word "race," for example, can't begin to capture the commonalities and differences of our shared history. Most African Americans have European ancestors; all European Americans have African ancestors. "Race" disguises rather than acknowledges our multifaceted histories.

Second, we have to remember how small the genetic differences among groups are. The genetic variants affecting skin color and facial features probably involve a few hundred of the billions of nucleotides in a person's DNA — an insignificant amount. Yet societies have built elaborate systems of privilege and control around these minuscule genetic differences.

Finally, we must become much better at putting genetics in context. People tend to attribute great importance to the findings of geneticists. But the striking homogeneity of our DNA actually emphasizes the centrality of individual and group experience in determining who we are. Everyone is the product of a particular human and genetic history. Yet this history is shared as well as unique, universal as well as individual. As we learn more about our relationship to the past, we need to find ways of interpreting this information that don't constrain the human spirit.

II

The Middle East

Encounters with the Other

Modern Humans and Neandertals in the Middle East

> When I look back upon my last months in Africa, it seems to me that the lifeless things were aware of my departure a long time before I was so myself. The hills, the forests, plains and rivers, the wind, all knew that we were to part.
>
> — Isak Dinesen, *Out of Africa*

SKHUL IS NOT a large cave. Its circular mouth, perched above a ledge of rock partway up a limestone cliff, is perhaps twice the height of a man, and its interior extends just a couple of strides into the cool, shadowy earth. It was springtime when I visited, and the entrance to the cave was garnished with wildflowers — thistles, lavender, wild chrysanthemums. From the cave's threshold I looked across a small stream crowded with eucalyptus trees to a steep, rocky ridge. To the west, beyond a banana plantation and the highway carrying traffic from Tel Aviv to Haifa, the sun glinted off the gentle swells of the Mediterranean. I was completely alone. The only sounds were the wind rising through the wadi and the gentle whine of the distant highway.

About 100,000 years ago, a small group of humans stood on the ledge in front of this cave, gathered around a shallow hole that they had dug. One of them laid the body of a middle-aged man in the hole. They tucked his legs against his chest and positioned his arms on top of his body. In his hands they placed the jaw of a wild boar — a hunting trophy perhaps. Then they covered him with dirt and walked away.

On May 2, 1932, a team of Palestinian excavators under the direction of the British archaeologist Dorothy Garrod and her young

American assistant Theodore McCown uncovered the top of the man's skull, and then his arms and the jaw of the boar. In the weeks and months to come, more skeletons emerged from the hard-packed breccia in front of the cave. By the end of 1934 the team had unearthed the remains of eleven individuals from this "most ancient prehistoric necropolis," as McCown put it.

The bones uncovered at Skhul, one of the largest collections of prehistoric human fossils ever found, raised far more questions than they answered. Garrod also had been excavating at a much larger cave named Tabun a couple of hundred yards down the wadi from Skhul. There she had found fossils that were clearly the remains of Neandertals. The limb bones were short and thick, an adaptation to the cold that has gripped the Neandertals' European homeland for much of the past 500,000 years. The skull of a woman found in the cave had classic Neandertal features. Her forehead slanted sharply back from heavy brow ridges. The lower part of her face projected well forward of her wide nose.

The fossils from Skhul were different. The braincases were high and rounded. The limb bones were long, adapted to a warm climate. The cheekbones were concave, not flat or projecting as in the Neandertals. Though they still had some archaic features, such as brow ridges, the people who lived near Skhul looked much more like modern people than like Neandertals.

For many decades the relationship between the occupants of Skhul and Tabun remained mysterious. Did the modern-looking people occupy the Mideast after the Neandertals? If so, maybe Neandertals gradually evolved into modern humans, which might help explain why Neandertal fossils younger than about 30,000 years have never been found. Or had the modern humans come into the Mideast before the Neandertals, as some of the other archaeological evidence indicated? That might fit with the climatic history of Europe. Perhaps Neandertals had moved into the Mideast during the worst part of the last Ice Age to escape the relentless advance of the northern glaciers.

The questions remained unanswered until the 1980s, when two newly developed dating techniques — thermoluminescence and electron spin resonance — were applied to the fossils from Skhul and Tabun. The totally unexpected results were that Neandertals seem to have occupied the site both before and after modern humans. According to the new dates, Neandertals were living in Tabun as early as

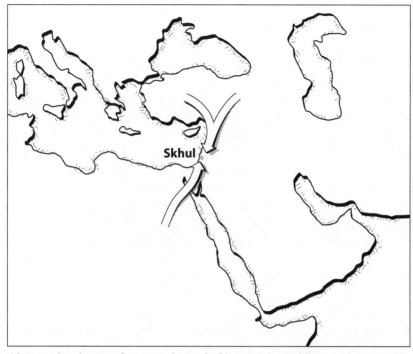

When modern humans first moved out of Africa into the Middle East about 100,000 years ago, they encountered Neandertals at the southern edge of Eurasia. One place where the two groups almost certainly came into contact was Skhul cave near Mt. Carmel in modern-day Israel.

200,000 years ago and as recently as 45,000 years ago, though it was impossible to tell if the cave had been occupied continuously.

Even more shocking were the dates for the Skhul fossils. Before the 1980s, most paleoanthropologists had assumed that the burials at Skhul took place 40,000 to 50,000 years ago, shortly before modern-looking people appeared in Europe for the first time. The new date was 100,000 years, double that estimate. The fossils from Skhul, paleo-anthropologists now realized — along with similar fossils from a cave named Qafzeh near the town of Nazareth — represented the earliest modern humans ever found outside of Africa.

Discussions of the interactions between Neandertals and modern humans usually center on Europe, where modern humans clearly had a more advanced culture than did their archaic relatives. But the two groups encountered each other first in the Middle East, and there they were much more evenly matched. In fact, the Neandertals prevailed in the Middle East for many thousands of years. By 80,000 years ago the populations of modern humans that had been living in Skhul and Qafzeh either had gone extinct or had retreated back into Africa. But about 45,000 years ago modern humans reappeared in the Middle East, and this time the Neandertals gave way.

The interactions between modern humans and Neandertals in the Middle East remain mysterious in many ways. They may have been enemies fighting to the death over disputed land. Or they may have exchanged food, stone tools, and perhaps even mates. These issues have a relevance that extends far beyond the shores of the ancient Mediterranean. They involve the origin of modern language, which may well have happened in or near the Middle East. They relate to whether modern humans mated only among themselves or whether they mixed with the archaic humans they encountered outside Africa. The ways in which these groups interacted in the Middle East can tell us much about what it means to be human today.

If Neandertals hadn't existed, writers would have had to invent them. The image of a primitive man peering through the cracked mirror of civilization is a staple of literature, from Jacob's brother Esau in the Book of Genesis (Esau is a "skilled hunter" covered with hair) to the villain of Robert Louis Stevenson's story "The Strange Case of Dr. Jekyll and Mr. Hyde." As soon as the fossils of Neandertals were discovered, they proved the perfect stand-in for the dark, brutish side of human nature.

The first fossils recognized as Neandertals were found in August 1856. Two quarrymen were shoveling debris from a limestone cave near Dusseldorf, Germany, when their shovels struck bone. Gradually they uncovered a pelvis, a skull with powerful, arching brow ridges, and other parts of a skeleton. They showed the bones to their foreman, who thought they were the remains of a cave bear. Still, he had the foresight to set the bones aside for a local schoolteacher known to be an amateur natural historian. The teacher knew immediately what they were: the bones of a human unlike any living today.

Paleoanthropologists are strict about priority in naming species. The quarrymen were digging in a cave in the Neander Valley. (In the nineteenth century, the German word for valley was *thal*, but the spelling was changed to *tal* at the beginning of the twentieth century, since German does not have a *th* sound.) The valley was named after a seventeenth-century composer and poet named Joachim Neumann (Newman in English), who signed his compositions with the Greek version of his name, Neander. Thus the irony of Neandertal man's literal translation: "man of the valley of the new man."

The timing of the discovery could not have been better. Three years later Charles Darwin, in his book *On the Origin of Species by Means of Natural Selection*, broached the unthinkable. Maybe humans weren't created a few thousand years ago by a benevolent God but instead were descended from earlier forms of life, including primitive cavemen. The idea was as horrifying to Victorian England as it is to many Christian fundamentalists today. As the bishop of Worcester's wife famously exclaimed, "Descended from the apes, my dear? Let us hope that it is not true, but if it is, let us pray that it will not become generally known."

Neandertals quickly proved irresistible to fiction writers. Authors from Arthur Conan Doyle to Isaac Asimov have spun tales around encounters between Neandertals and modern humans, though their themes have usually had more to do with the concerns of the day than with established fact. In his 1921 story "The Grisly Folk," for example, H. G. Wells describes the initial encounters of Neandertals with modern humans, the "early pilgrim fathers of mankind." With their upright bearing and noble demeanor, the invading moderns seem every bit the intellectual equals of the conquering European colonists of the nineteenth century. The Neandertals, on the other hand, are "ugly, strong, ungainly." They "shambled along with a peculiar slouch, they could not turn their heads up to the sky." The Neandertals steal a modern child, presumably to roast and eat it, and they kill one of the leaders of the moderns. When the story was written just after World War I, the descent into mutual savagery must have seemed inevitable. "For the Neandertalers it was the beginning of an incessant war that could end only in extermination," writes Wells. "By ones and twos and threes therefore the grisly folk were beset and slain, until there were no more of them left in the world."

By the 1970s, impressions of the Neandertals had softened. In her mega-bestseller *The Clan of the Cave Bear*, Jean Auel presents a much more sympathetic portrait of these early humans. In the second chapter, Auel describes one of the Neandertal women this way: "She was just over four and a half feet tall, large boned, stocky, and bow-legged, but walked upright on strong muscular legs and flat bare feet. Her arms, long in proportion to her body, were bowed like her legs. She had a large beaky nose, a prognathous jaw jutting out like a muzzle, and no chin. Her low forehead sloped back into a long, large head, resting on a short, thick neck. . . . Big, round, intelligent, dark brown eyes were deep set below overhanging brow ridges."

In the novel the Neandertals adopt a five-year-old modern human child left orphaned and alone by an earthquake in the mountainous region north of the Black Sea. Given the name Ayla, the child is considered hideously ugly by the clan as she grows up. In fact, she is tall, blond, and blue-eyed, with full breasts and rounded hips — a beauty in contrast to the Neandertal beasts.

She is raped and impregnated by a Neandertal man. The son she bears is considered deformed by the clan. He has brow ridges but a straight, high forehead. His nose and jaws are much smaller than his father's. Below his mouth is "a bony protrusion disfiguring his face, a well-developed, slightly receding chin."

In the book's climactic scene, Ayla accidentally observes the secret rituals of the Neandertals' medicine men, and she links telepathically with the mind of the most powerful shaman (this is fiction, after all). At that moment the ultimate fate of the clan is revealed. "[This] race of men with great brains but no frontal lobes, who made no great strides forward, who made almost no progress in nearly a hundred thousand years, was doomed to go the way of the woolly mammoth and the great cave bear. They didn't know it, but their days on earth were numbered."

In a rare confluence of the popular and the scientific, the interests of fiction writers and those of anthropologists almost exactly coincide. To put it succinctly, those interests are violence, language, and sex.

No recent paleoanthropological debate has generated more acrimony than whether Neandertals and modern humans had sex. For years one group of paleoanthropologists insisted that Neandertals were the ancestors of today's Europeans. According to this view, archaic humans around the world evolved into modern humans, with just enough mating between the groups to keep them evolving in the same direction. Fossil and genetic evidence has severely undermined this view, and few paleoanthropologists still adhere to it fully. But a modified version has emerged, which posits that as modern humans came out of Africa, they mated with the archaic people they met, first in the Middle East and then elsewhere. As a result, segments of DNA from those archaic people still exist and influence the characteristics of people living today.

There's one sure way to prove that current humans have Neandertal ancestors: find some mitochondrial DNA from a Neandertal in a person living today. Such DNA should be easy to identify. According to the fossil evidence, the Neandertals evolved from a subbranch of archaic humans who made their way from Africa through the Middle East to Europe more than 500,000 years ago — well before the times when mitochondrial Eve or the Adam of the Y chromosome lived. In Europe the Neandertals must have been genetically isolated from Africans, or their features would not have diverged so dramatically. This means that their DNA, including their mitochondrial DNA, was also diverging. If Neandertal women later mated with modern men coming out of Africa, this distinctive mitochondrial DNA would enter some populations of modern humans. And if such mating was common, some of these Neandertal mitochondria should survive in people today.

No such DNA has ever been found, nor is it likely ever to be found. Of the many thousands of samples of mitochondrial DNA from around the world that have been sequenced, all are descended from mitochondrial Eve. The statistical chance that an aberrant mitochondrial sequence will be found is now very low. The same is true for all the Y chromosomes ever sampled; all come from the Y chromosome's Adam.

Still, the extinction of Neandertal mitochondrial DNA and Y chromosomes does not prove conclusively that Neandertals did not mate with modern humans. If, for example, Neandertals and modern humans occasionally mated, one of Eve's mitochondria might have found its way into the Neandertal gene pool relatively late in that group's history. Then Europeans still could have gotten their mitochondrial DNA from a Neandertal, even if it originally came from Eve.

Until recently, there seemed only one way to test this hypothesis. Build a time machine, travel back 40,000 years, and ask a Neandertal for a blood sample. Fortunately, an easier way now exists. When an organism dies, its DNA does not instantly turn to dust. It persists in the remains of the organism more or less intact until the combined effects of time and decay gradually degrade it. DNA in fossilized bones can remain in pretty good shape for thousands of years, especially if the bones are in a cool, dry place.

The study of ancient DNA has been extremely contentious. Early claims of successful sequencing turned out to be overblown. Contamination is a severe problem: the DNA of people handling a fossil may be transferred to the bones and then turn up in the DNA sequencer. And the older the DNA, the more it disintegrates. The re-creation of dinosaurs from the DNA retained in the guts of fossilized mosquitoes will remain safely in the realm of science fiction.

In the early 1990s the prospects of obtaining DNA from Neandertal bones seemed little better than that. But as laboratory techniques continued to improve, two young German researchers, archaeologist Ralf Schmitz and geneticist Svante Pääbo, became increasingly confident that they could obtain mitochondrial DNA from Neandertal fossils. Furthermore, they didn't want to test just any old Neandertal. They wanted to sample the very bones found by the quarrymen in 1856. Needless to say, the curators of the bones were not eager to sacrifice parts of their irreplaceable fossils to an unproven technique. "It was like getting permission to cut into the Mona Lisa," recalls Schmitz. But he and Pääbo persisted. Schmitz cajoled the museum curators, emphasizing the publicity that a positive find would generate. Pääbo demonstrated techniques he had developed in his University of Munich laboratory that offered at least some prospect for success.

Finally the museum agreed. In 1996 a professional bone preparer carefully sawed a half-inch-wide, eighth-of-an-ounce chunk from Neandertal man's right upper arm bone. Pääbo took the plastic vial containing the bone sample back to his lab and handed it to graduate student Matthias Krings, who had just spent nearly three years trying unsuccessfully to extract DNA samples from Egyptian mummies. "At the beginning I didn't have much hope that it would work," says Krings, who is now employed by a biotechnology consulting firm in Munich. "I was having trouble getting DNA from samples that were 5,000 years old. Now I had a sample that was at least 35,000 years old, and maybe more than 100,000 years old."

Krings ground up a small section of the bone, mixed the powder with a liquid solution, and removed the DNA-containing portion using a centrifuge. He then mixed the DNA with what are called molecular primers. These primers latch onto specific regions of DNA — in this case, two twenty-base-pair segments known to exist in the mitochondrial DNA of modern humans and thought likely to exist in Neandertals. By adding the proper compounds, one can copy all of the DNA between a pair of primers, yielding twice as much DNA as in the original sample. The cycle is then repeated. After two cycles, the DNA has increased four times; after three cycles, eight times; and so on. In this way, even very tiny fragments of DNA can be copied many times, yielding enough DNA for sequencing.

My description makes the process sound easy. In fact, it's devilishly hard. Coaxing DNA that is tens of thousands of years old from tiny fragments of bones requires endless care and hundreds of separate laboratory manipulations. Krings put in one-hundred-hour weeks for months on end.

But then he began to find fragments of DNA that were clearly different from the DNA of today's humans. At position 16,223 in the mitochondrial sequence, where most humans have a cytosine, the sample from the bone had a thymine. At 16,254, instead of the usual guanine, the sample had an adenine. "I can still remember the feeling I had," Krings says. "By this time I was able to spot a difference if I saw one, and when I saw differences in the extraction, I knew we had something." Using different primers, Krings was able to piece together a mitochondrial DNA sequence of 379 nucleotides from the fossil. This sequence differed from the standard modern human sequence at twenty-seven locations — far too many for the DNA to have come from mitochondrial Eve.

Krings couldn't celebrate yet. He needed an independent confirmation of the results. He sent a separate sample of the bone to Mark Stoneking, then at Pennsylvania State University, and Stoneking passed it on to graduate student Anne Stone. Now an assistant professor at the University of New Mexico, Stone had just finished her Ph.D. and still had a few months before graduation, so she agreed to do the project. But it was much more difficult than she had expected. "Even grinding the bone into powder was hard," she says. "I had to drill it full of holes, like Swiss cheese." Not until the day of graduation did the results finally begin to emerge from the sequencer. While her parents waited in her apartment, Stone ran back and forth to the lab. "I called Germany and began to read the differences to Matthias," she recalls. "I'd say, 'I got this base pair at this position,' and he would cheer. Then I'd say, 'I got this base pair at this position,' and he would cheer again."

"I can tell you," says Krings, "we had a party that night." When the paper describing the sequence was published a few months later, one paleoanthropologist called it "as exciting as the Mars landing."

Using the differences in sequences, Krings and Pääbo calculated how long the Neandertals must have been genetically separated from our African ancestors. The results fit the fossil record perfectly. For their mitochondrial DNA to be so distinct, the lineages leading to Neandertals and to modern humans must have been separated for about 500,000 years.

The absence of Neandertal mitochondrial DNA in modern humans greatly reduces the possibility that interbreeding occurred — a result that has been confirmed several more times by the sequencing of mitochondrial DNA from other Neandertal bones. Still, the case is not airtight. As I've said, mitochondrial DNA makes up just a tiny fraction of our total DNA. Perhaps pieces of Neandertal DNA persist in the chromosomes of people today, even if they have been lost from the mitochondria.

But that case, too, was made less likely with the publication of a study by a team of German researchers who compared chromosomal DNA from a Neandertal with modern human DNA. In general, chromosomal DNA in fossils is much more difficult to sequence because each cell contains many mitochondria but only a single copy of the chromosomes. But samples of chromosomal DNA can be compared another way. When two DNA samples are mixed, they tend to stick together. The greater the similarity of the two sequences, the greater the stickiness. Researchers used this technique to compare chromosomal DNA from a person living today with three different samples of ancient DNA: one from a 35,000-year-old modern human fossil, one from a Neandertal who lived 50,000 years ago, and another from a 100,000-year-old Neandertal fossil. The DNA from the modern human was much more similar to today's DNA than it was to either of the Neandertals.

None of these results surprised geneticists. For years they have been looking for the DNA of archaic humans in modern people — without success. If such DNA exists, it shouldn't be that elusive. If Neandertals were among the ancestors of modern people, then the DNA of Middle Easterners and Europeans should display unique genetic sequences, because Neandertals lived only in the Middle East and western Eurasia. But no such unique sequences show up. The DNA of everyone on the planet today is pretty much the same, as would be expected if we all had the same origins.

The lack of evidence for interbreeding between modern humans and Neandertals is a mystery. Modern humans — well, males at least — will copulate with almost anything. According to data gathered by Alfred Kinsey and his colleagues in the 1940s, 8 percent of American men reported having sexual experiences with animals. Among men who grew up on farms, the percentage was much higher — approaching 50 percent. As modern human males began encountering Neandertal women in the Middle East and Europe, it seems inconceivable that they would not have mated with them, no matter how different they looked.

One possibility is that Neandertals and the ancestors of modern humans had been separated for so long that their DNA was incompatible. When a sperm cell from a modern human and an egg from a Neandertal (or vice versa) fused, their chromosomes might not have lined up, making reproduction impossible. Or the chemical signals on the egg might not have allowed sperm from members of the other group to penetrate the cell wall. Some anthropologists have suggested that Neandertal children remained in the womb for longer than nine months — perhaps for a year — and that they matured faster after birth. Such distinctions would seem to indicate that modern humans had crossed the evolutionary Rubicon from Neandertals, that they were different species.

But the existence of reproductive barriers between the two groups is far from certain. Species that have been separated for much longer than Neandertals and modern humans can still interbreed. The two separate chimpanzee species that live in equatorial Africa — the common chimp and the pygmy chimp or bonobo — have been separated genetically for more than 2 million years. Yet when a female chimp and a male bonobo are put together in a cage, they mate and have offspring. Lions and tigers will also mate, although not all of their offspring will be fertile, as will many other related species.

Another possibility is that hybrid human-Neandertal children were born but were shunned. If hybrids were so strange-looking that their families rejected them, the children would die and Neandertal genes would have no way of becoming part of the modern gene pool. But it seems unlikely that all such children would be abandoned. Perhaps communities of hybrids sprang up over the millennia — paleoanthropologists still argue fiercely over the characteristics of newly discovered fossils. But if hybrids did exist, their contributions to today's gene pool must have been very small.

Yet another possibility is that Neandertals and modern humans were making war, not love. If the two groups were locked in combat, few hybrids may have been born (unless women were captured or raped by the victors). However, no evidence exists of hostilities between the two groups. Neandertal skeletons bear no obvious signs of fighting, and modern human fossils with such injuries don't show up until a period when the moderns must have been fighting among themselves.

Humans certainly have the capacity for violence, and over many thousands of years of coexistence conflicts between moderns and Neandertals had to occur. But violence was not inevitable. For one thing, the image of modern humans marching north from Africa like colonial conquerors is obviously wrong. Most modern humans were probably scarcely aware that they were moving. With each new generation, they might have expanded their range by just a few miles — to escape overcrowding, perhaps, or to forage in new areas. Still, over many generations, such expansions could spread across thousands of miles.

Given what is known today, the most likely scenario is that modern humans gradually displaced their archaic neighbors through some technological or social advantage. The Neandertals could have been pushed onto difficult terrain where they could survive during periods of benign weather but not when the climate worsened. Even if the advantage of modern humans was very slight — gathering food a little more effectively or having somewhat more children — they could have replaced archaic humans over time. According to one computer model, if the mortality of Neandertals was just a couple of percentage points higher than for modern humans, the Neandertals could have been driven to extinction within a thousand years.

Why were modern humans ultimately successful in replacing the Neandertals? Paleoanthropologists and geneticists have proposed a number of possibilities, such as greater mobility or better planning abilities, but the favorite has always been language. If modern humans came out of Africa happily chattering away — so the story goes — they could easily have overrun archaic humans who communicated at best in grunts.

However, this scenario has a big problem. Say that the modern humans at Skhul 100,000 years ago had language and their Neandertal neighbors at Tabun did not. In that case, the behaviors of the two groups should have been quite different. The moderns should have had better tools, more extensive trading networks, and different ways of procuring food. Their culture should have been more sophisticated — and that sophistication should be apparent in the artifacts found with their fossils.

It isn't. On the contrary, the moderns at Skhul seem to have behaved almost exactly like the Neandertals down the valley at Tabun. They made the same stone tools. They hunted the same animals. True, the moderns seem to have buried some of their dead, but Neandertals may have done that, too.

The issue then becomes, were both groups using language, or neither? This question bears on another extremely contentious topic in paleoanthropology. The traditional view is that language arose very early in human history, during the origins of *Homo* about 2 million years ago. Brain size increased dramatically in the initial species of *Homo*, and a logical conclusion is that all that extra gray matter was being used for language. For one thing, the production and use of stone tools — *Homo*'s major distinguishing characteristic — would seem to require language. A stone knapper needs to know the shape of the tool he or she is to craft, and this understanding must be passed from one generation to the next. Both tasks seem complicated enough to demand the kind of symbolic knowledge embodied in language.

Yet if archaic humans did have language, what a strange language it must have been. Once a particular variety of stone tool appears in the fossil record, it remains virtually unchanged for many thousands of years. Neandertals made more or less the same scrapers and spear points for more than 5,000 generations.

How could humans have had language, with all its playfulness and variety, and yet have endured lives of such unimaginable technological monotony? When humans give names to things, the idea embodied by that name itself becomes a tool to be used. People think of different uses for that tool or modify it slightly to serve new purposes. Once the name of something is severed from its physical embodiment, people can manipulate that thing in their minds.

Perhaps Neandertals and other archaic humans did not have language — or at least not the kind of language we have today. Studies of chimpanzees show that quite sophisticated behaviors can develop even without language. Chimps use rocks to crack open nuts and probe termite holes using sticks, and they are able to pass on cultural innovations to their offspring. Of course, Neandertals lived much more complicated lives than do chimps — they hunted large animals, tended fires, and stored food for the winter. Yet nothing in their lives seems to have required language. They did what they had to do to survive, what they had always done. They did not think about how to do things differently to make their lives better.

The Neandertals were smart — their brains were, on average, larger than those of people living today — yet they seem to have devoted their considerable intellect to living in the moment. They seem not to have developed the fluent language that lets us wonder, adapt, and create.

Between 100,000 and 65,000 years ago, something remarkable happened in human history. When modern humans again began to move out of Africa, they demonstrated a very different culture. Their stone tools were smaller and more sophisticated. They worked bone and antler into carefully predetermined shapes. They wore jewelry made from shells, bones, and animal teeth. They covered their dead with red ocher and buried them in graves. Most telling of all, they created art, something humans had never done in the past.

Not all of these behaviors appeared full-blown in the Middle East. The earliest clear evidence for modern behaviors comes from Australia, which modern humans apparently reached as early as 65,000 years ago. Modern tools may have appeared first (the dating is uncertain) in a broad arc extending from eastern Africa to eastern Europe, and the most impressive surviving pieces of early artwork come from France and Spain. But the modern humans who began moving into the Middle East about 45,000 years ago were obviously distinct culturally from their Neandertal predecessors. Their tools were much more sophisticated, varied, and changeable. They developed regional tool-making styles, sweeping away the monotony of previous millennia.

Cultures of such sophistication and adaptability could not have existed without the kinds of languages we use today. People were exchanging goods and ideas. They obviously had complex conceptions of themselves and the physical world, as represented in their art and tools. The contrast with the preceding period of human existence could not be sharper. Perhaps the Neandertals had some sort of proto-language based on simple nouns and verbs, like the pidgins that develop today among speakers of different languages. But they and early modern humans seem not to have had language as we know it today. Rather, language seems to have appeared sometime between the initial unsuccessful migration of modern humans out of Africa 100,000 years ago and their successful expansion to Australia 65,000 years ago.

Biologists tend to look for some sort of genetic trigger for language. The simplest form of the argument goes like this: Maybe a mutation occurred that allowed an individual to talk. The mutation might have caused the brain to grow in a different way or forged internal connections that weren't there before. Because that mutation enabled the person who had it to reproduce more successfully, it spread throughout the population via natural selection.

But such explanations for the appearance of language have always been suspect. The abilities of the human mind depend on many genes working together. Those abilities emerged over time from a multitude of genetic changes as various species of first *Australopithecus* and then *Homo* evolved. The ability to speak could not depend on a single gene, because if the function of that gene were lost through a subsequent mutation then language would be lost as well.

Besides, if language really originated in the past 100,000 years, the "biological trigger" hypothesis must be wrong. The basic body plan of modern humans was already established 100,000 years ago as modern humans were spreading through Africa. The biological changes necessary for language would have had to spread very quickly and thoroughly through a widely dispersed population, which is genetically implausible.

The real explanation may be much simpler. Maybe language did not appear immediately after some evolutionary threshold was crossed. Maybe it was invented. At some point a group of humans may have discovered that objects, actions, and feelings could be assigned arbitrary sounds. They then could call upon their existing vocal apparatus to make those sounds.

Of course, language could not have come from nowhere. To speak, early humans needed particular vocal and neural mechanisms. But here a notorious problem arises. Any adaptations produced by evolution are useful only in the present, not in some vaguely defined future. So the vocal anatomy and neural circuits needed for language could not have arisen for something that did not yet exist. They must have had some additional purpose.

Brain scientists have proposed some ideas. Archaic humans undoubtedly used their vocal cords to make sounds. If they used a simple protolanguage, natural selection may have favored anatomical adaptations useful for language. Also, many of the mental networks used for language are involved in other fine motor skills, such as throwing objects and pointing at things. Maybe selective pressures for these activities produced expansions in parts of the brain that were later used for speech.

Once people began using language, human societies must have changed irrevocably. Children would be taught from an early age how to make sounds and how to associate those sounds with things external to themselves. At that point there would be no going back to a protolanguage or to no language at all. Language would become part of being human.

Linguists have no idea exactly where language appeared. A good guess might be somewhere in eastern or northeastern Africa. Some tantalizing signals in the African archaeological record from the period between 100,000 and 65,000 years ago might be related to language. Bone points, including harpoon heads, controversially dated to 90,000 years ago, have been found in central Africa. Farther south, a collection of artifacts 50,000 to 70,000 years old looks remarkably modern. But these flashes of modernity were succeeded by the same old stone tools that had been used for millennia. Scientists do not know whether these bursts of creativity were the first signs of language appearing in modern humans or examples of inventions by people lacking the words necessary to disseminate their ideas.

The invention of language marked a fundamental change in human consciousness. We think in words. Indeed, the kinds of thinking we do would not be possible without words. None of us can directly experience the consciousness of a nonhuman animal because we cannot escape the consciousness defined by our use of language. Language did not just reveal an already existing consciousness to others; it created that consciousness. With the invention of language, the story of humanity truly begins.

The invention of language does not directly explain why modern humans seem not to have bred with Neandertals. Yet it highlights a difference between the two groups that, when combined with their differences in appearance, might make their separateness less mysterious. Modern humans seem to have a unique capacity and desire to learn new things. That doesn't mean the early moderns learned to speak — or learned anything else — in a single moment of illumination. Their new behaviors must have emerged in stages. But perhaps that is our defining characteristic. Maybe we are the first creatures on earth driven by a need to know.

One last glimpse at the Neandertals reveals both the tragedy and the inevitability of their demise. In a cluster of limestone caves southeast of Paris, archaeologists have found a distinctive assembly of modern-looking tools, beads, and ornamental objects dated to about 36,000 years ago. Initially, scientists assumed that these artifacts, which belong to a culture known as the Chatelperronian, were the work of early modern humans in Europe. But then fossils were found with a Chatelperronian deposit, and they were clearly those of Neandertals.

Whatever their other limitations, Neandertals apparently were able to learn. They watched the modern humans who were growing more numerous all around them. They copied the tools and other artifacts that seemed to give the moderns an advantage. Perhaps they learned to speak the moderns' language.

We will never know how much the Neandertals were like us. My suspicion is that they had more of our abilities and desires than the common caricatures of them would suggest. But by 30,000 years ago their time was past. The future belonged to the new people coming out of Africa.

5 ➤➤

Agriculture, Civilization, and the Emergence of Ethnicity

> I pledge allegiance to the Flag
> Of the United States of America
> And to the Republic for which it stands
> One nation, under God
> Indivisible
> With liberty and justice for all.
>
> — U.S. Pledge of Allegiance

EAST OF JERUSALEM, just past the new homes erected by Israeli settlers, Highway 30 heads steeply downhill. Across barren hillsides and bone-dry wadis it descends until your ears pop with the pressure change. A concrete pillar announces that you've reached sea level, and still the road drops lower. A line of hills appears on the eastern horizon — you're driving into a valley with a broad, flat floor. Soon the road levels out and reaches an intersection. To the right is the Dead Sea, its lifeless shoreline rimmed with salt. To the left, just a few miles up the valley of the Jordan River, is the Arabic town of Riha, also known as Jericho.

Today little distinguishes Jericho from the other Palestinian towns scattered throughout Israel and the West Bank. Trucks fill the air with gasoline fumes as farmers unload vegetables at markets. Up and down the street men in sidewalk cafés drink coffee and read Arabic newspapers. The only sign of Jericho's storied past is an odd mound of dirt, perhaps 25 yards high and 250 yards across, perched at the northeast corner of the town, next to a spring that feeds water to the surrounding fields. According to the Hebrew Bible, this mound is where the ancient Israelites launched their conquest of the land of Canaan. They had wandered in the deserts south of the Jordan Valley for forty years after Moses led them out of Egypt. Camped amid the hills to the east of Jericho, they looked down on the green fields and palm trees of the valley and knew they had reached the promised land. From the top of one of those hills, God allowed Moses a glimpse of the territory he had promised to his people. Then Moses died and was buried in a place that no one knows "unto this day."

Under the leadership of Joshua, the Israelites crossed the Jordan and halted before the walled city of Jericho. For six days the Israelite soldiers marched around the city behind the ark bearing God's covenant with the Hebrews. On the seventh day, seven priests at the head of the marching warriors blew on seven ram's horns. The people raised a great cry, and the walls of the city fell down flat. Then, as the Hebrew Bible says, "they devoted to destruction by the edge of the sword all in the city, both men and women, young and old, oxen, sheep, and donkeys."

Since the nineteenth century, many archaeologists have gone to Jericho to find the ruins of the ancient city. When they dug into the mound, they discovered that it was not an odd quirk of the valley's terrain. It was the accumulated remains of thousands of years of ruined houses, temples, graves, roadways, and garbage. They dug trenches in the mound and found grinding stones, jewelry, pottery, and in great profusion the remains of collapsed walls. Surely a great city once stood here.

One of the archaeologists who went to Jericho was Kathleen Kenyon, whose excavations in the Middle East made her one of the most famous scientists of her generation. She arrived in Jericho in 1953, intending to stay for just a year or two to complete some work left undone by a colleague. She stayed for eight years and learned more about the city than anyone else ever has. Like many of her predecessors, she promoted her work by hinting that she was uncovering proof of the literal accuracy of the Bible. "One can visualise the Children of Israel marching round the eight acres of the town and striking terror into the heart of the inhabitants," she wrote in her 1957 book *Digging Up Jericho.* Perhaps the shards of pottery found in an abandoned room were those of a Canaanite woman "who may have dropped the juglet beside the oven and fled at the sound of the trumpets of Joshua's men."

But Kenyon's biblical allusions always seemed a bit forced. In the end she concluded that Jericho was probably already in ruins during the period recounted in the Hebrew Bible. She was much more interested in what came before the biblical era — not 3,000 years ago, but 10,000.

In the Middle East, history exists as a kind of ghostly superposition, like a piece of film that has undergone multiple exposures. According to the New Testament, Jesus walked past Jericho to be baptized in the Jordan River. Thousands of years before biblical times, some of the first people ever to live in permanent homes drew water from Jericho's spring. More than a million years ago, an early species of *Homo* lived in this valley, which seems appropriate, given that the Jordan Valley is the northernmost extension of the Great Rift Valley of Africa, where so much of human evolution took place.

Jericho also figures in what is without doubt the most important development in human history after the invention of language: the emergence of agriculture, about 10,000 years ago. From today's perspective, planting seeds and raising animals would seem to be obviously better ways to acquire food than foraging in the wild. But the advantages of agriculture were not at all obvious to hunter-gatherers. Farming required new ways of living and interacting with others. It made inordinate demands on the early farmers, who usually had to work much harder to feed themselves than did the hunter-gatherers in the surrounding hinterlands. Making the transition to agriculture took thousands of years for many groups of people, and some groups never made the change at all.

The invention of agriculture transformed human life. The domestication of plants and animals led to rapid expansions of populations. Greater population densities helped produce cities, warfare, nations, and mass religions. Urban civilizations changed how people choose their mates and created new kinds of social structures, including what we now call ethnic groups. The seeds of the modern world were planted at Jericho, and we cannot understand important features of the world today without looking at the origins of farming, civilization, and ethnicity in the Middle East.

*

When modern humans from Africa reentered the Middle East about 45,000 years ago, their material culture was clearly more sophisticated than that of the Neandertals. They made slender, delicate stone blades. Worked bone and antler also appear in the archaeological record, along with grinding tools and shell jewelry.

Yet for all their new abilities, for the next 35,000 years the modern humans in the Middle East relied largely on the same resources the Neandertals had. They gathered the fruits and seeds that grow in the region — pistachios, almonds, chickpeas. They hunted wild animals, especially gazelles and deer, in the rugged terrain bordering the Mediterranean. They were continually on the move, following the seasons and migrations of game.

Anthropologists have two ways of learning about the daily lives of our prehistoric ancestors. One is by studying the artifacts they left behind — mostly stone tools and the remains of structures. This is an immensely difficult task. Imagine trying to reconstruct modern society after some sort of bizarre atomic bomb has destroyed everything except our kitchen utensils. Given how little they have to work with, archaeologists have learned an incredible amount about the past, though inevitably their interpretations sometimes overstep the bounds of the evidence.

The other way to learn about ancient humans is to study people presumed to be living in more or less the same way today, such as the Bushmen. But these extrapolations must be made with great caution. No one knows if ancient people had the same social structures or behaviors as modern foragers, who almost always interact with agriculturalists. Moreover, the social structures of today's hunter-gatherers vary widely, and preagricultural societies may have been no less diverse.

Still, a couple of fairly reliable generalizations can be made. At the most basic social level, hunter-gatherers tend to organize themselves into groups that anthropologists traditionally have called bands. These consist of one or a few closely related families, probably ten to fifty people altogether. They tend to be fairly egalitarian, though a dominant family member may act as leader. If one were to identify a group comparable to a band in an industrialized society, the closest analogue would be the family members who gather at someone's home for a celebration or holiday. (This comparison suggests a somewhat alarming thought experiment. Imagine living with those family members all the time, in makeshift shelters, gathering food from the wild, and rarely seeing other human beings. As should be obvious, anthropologists have no reason to think that life in bands was always peaceful.)

Theoretically, a band of humans can be completely self-sufficient. By mating within the group and staying out of the way of other humans, a band can eke out a marginal, if monotonous, existence. But such groups have many good reasons for interacting with others. They can trade tools, food, or ideas. They can create webs of mutual obligation, so that one band can rely on another during times of crisis. Bands can forge alliances to protect territory, resources, and families.

Most important, bands can exchange mates. A small group can certainly mate only within itself, as has occurred occasionally in the past. But most people are not eager, whether for social or biological reasons, to mate with close relatives. If older men mate with younger women, the younger men are left out in the cold. And prolonged periods of intermarriage within very small groups may cause a small but significant rise in genetic problems among offspring (although a certain degree of inbreeding may not be as risky as is commonly thought).

Because of the benefits to be gained by interacting, bands throughout history have tended to organize themselves into larger units, which anthropologists traditionally have labeled tribes. Today anthropologists often shun the term because it connotes a particular kind of society, and the social arrangements between bands are in fact extremely varied. Still, I'm going to use the word "tribe" with the proviso that it can describe a broad range of social structures.

One important aspect of tribal life is that everyone has at least the potential of knowing everyone else personally. This typically means that the tribe is limited to several hundred members, though some groups are substantially larger. Most members are related to one another through blood or marriage, though the relation may be a distant one. The bands within a tribe may live close to each other, or they may gather at a central location periodically to exchange goods and to intermarry. All members of a tribe usually speak the same language, though exceptions occur. More broadly, they typically share a way of life and a belief system — they are united by culture.

For most of the history of modern humans — from our appearance more than 100,000 years ago until the invention of agriculture — human social life took place entirely in the context of bands and tribes. The patterns of genetic variation that exist in the world today therefore were heavily influenced by how bands and tribes interacted. If matings occurred between two different tribes on a regular basis, they became more genetically similar. If tribes were isolated from one another — by geographical barriers such as deserts, mountains, or oceans or by cultural barriers such as language or religion — their genetic patterns diverged.

Most of the physical differences between today's human groups probably arose during the exclusively tribal phases of our history. Geographic barriers would have presented obstacles to gene flow. And even today, tribes (or groups that think of themselves in tribal terms) tend to practice endogamy — that is, the members usually marry within the tribe — and view other tribes with suspicion. If these attitudes were common in the past, they could have severely reduced the genetic exchanges among groups of people. Indeed, if such attitudes shaped the behaviors of the earliest modern humans, they could help explain the lack of interbreeding between modern humans and archaic humans.

At the same time, the tribal divisions of modern humans clearly have not been inviolable. The haplotypes of mitochondrial DNA and the Y chromosome are too mixed for groups to have been strictly separated. The haplotypes found in the Middle East are a good example of this mixing. The modern humans who began entering the Middle East 45,000 years ago must have come from the northeastern corner of Africa, and they would have brought with them a subset of the mitochondrial and Y haplotypes then existing in Africa. Many of these haplotypes can still be found in the Middle East among individuals who trace the bulk of their ancestry to people who lived in that region. At the same time, some of the descendants of the first modern humans in the Middle East expanded into Europe and the rest of Asia, where they formed new and distinct human groups. Over time new mutations occurred in the DNA of these groups, creating new haplotypes. Later some of the members of these groups returned to the Middle East, where they mixed with people whose ancestors had never left. These complex migrations have produced, in the Middle East today, one of the most diverse collections of human DNA found anywhere in the world.

Beginning about 12,000 years ago, hunter-gatherers near the eastern Mediterranean shoreline began to adopt a new way of life. Instead of continually moving from place to place, they began to live in single places for much or all of the year. They learned to exploit the wild plants and animals in an area much more intensively. For example, they embedded tiny stone blades into wooden or bone shafts to make sickles, which they used to harvest the wild grasses around their homes. They made seasonal trips to hunting camps and returned with their catch to base camps, where they built permanent structures — rounded rooms dug partly into the ground with roofs probably made of brush.

Once people became more sedentary, their lives changed. They no longer had to haul all their possessions from one camp to another. As a result, they immediately acquired more stuff (as we would say today). The remains of their settlements include heavy stone mortars used to grind grain. They made cups and bowls and stored goods in pits lined with stone slabs. The first tools belonging to this tradition were found at a mound around a spring called Wadi Al Natuf just north of Jerusalem, and this period in Middle Eastern history is called the Natufian.

The earliest archaeological remains at Jericho, at the base of the mound, were left by the Natufians, though they seem to have used Jericho only as a temporary camp. Then the area around the spring was uninhabited for a long period. The archaeological remains from the next settlement at the spring, dated to about 10,000 years ago, aren't those of a camp. They're the remains of a town large enough to hold several hundred people.

These first inhabitants of Jericho were well on the way to full-fledged agriculture, though they continued to rely heavily on the wild plants and animals of the region. They were experimenting with sowing the seeds of chickpeas, lentils, wheat, and barley near the spring. Animal bones found at the site indicate that the people may have been domesticating livestock, or at least importing domesticated animals from elsewhere.

Kathleen Kenyon often called Jericho the "world's oldest town," and civic boosters in Jericho still make that claim. Actually, a number of similar towns existed about the same time up and down the Jordan Valley, along the Mediterranean coast, and north into Turkey. Still, Jericho was one of the largest and most sophisticated of these towns, as well as one of the richest; rare obsidian, turquoise, and cowrie shells all have turned up in the ruins.

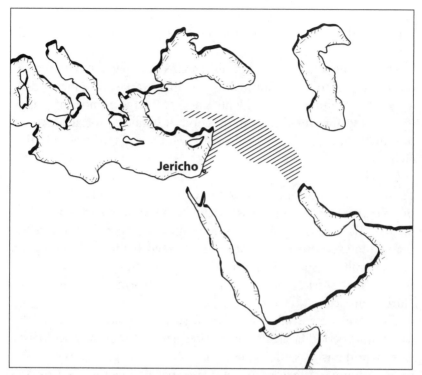

Agriculture first developed about 10,000 years ago in a broad arc of mixed terrain extending from the area near the Dead Sea north into the mountains of Anatolia and east along the foothills of the Zagros Mountains.

Just how much life had changed for the people of Jericho since hunter-gatherer days is apparent in a remarkable structure discovered by Kenyon. Sometime between 10,000 and 9,000 years ago, the occupants of Jericho built a wall around their town. It was 10 feet wide at the base, as much as 13 feet high, and 700 yards in circumference. Outside the wall was a moat that was 30 feet wide and 10 feet deep. At one end of the town, just inside the wall, a tower rose as high as a three-story building, with a stone staircase rising through its center. As Kenyon wrote, the tower would not disgrace "one of the more grandiose medieval castles."

Kenyon assumed that the purpose of the wall was to defend against marauding armies intent on acquiring the city's agricultural surpluses. But military fortifications don't appear around other towns for thousands of years, and the wall of Jericho may have had a different purpose. It might have been built to protect the town from flood-induced mud flows, which must have become especially severe after the people cut down the trees and harvested the brush on the surrounding hillsides.

Though the function of the wall remains unknown, it marks an innovation in history even more dramatic than warfare. It signals a dedication to communal life, the willingness of a large number of people to work together for the common good. It heralds a social organization that, within a few thousand years, would lead to the world's first civilizations.

The first people in the world to domesticate plants and animals were the occupants of towns such as Jericho in the western part of the Middle East. But elsewhere other groups of hunter-gatherers also were beginning to experiment with farming. Between 10,000 and 4,000 years ago, agricultural societies sprang up in eastern and southeastern Asia, New Guinea, Africa, Central America, South America, and eastern North America.

The development of agriculture in so many places within a relatively short period has led to much speculation among archaeologists. One hypothesis was that ancient peoples exchanged ideas through some sort of prehistoric grapevine. This speculation was often applied not just to agriculture but to pottery making, iron smelting, and other technologies as well. Usually it was couched in terms of a superior civilization bestowing the fruits of its culture on a benighted people too ignorant to invent anything for themselves.

But the more closely archaeologists looked at these so-called diffusionist theories, the more unrealistic they seemed. No evidence supports the idea that the various centers of agricultural development communicated with one another. Uninhabitable deserts, impassable oceans, and vast distances separated these regions. If widely separated ancient peoples had been in regular communication with each other, human history would have been quite different.

In the case of agriculture, other forces must have caused its independent development in numerous places around the world. The quest to identify these forces has long attracted scholars of a particularly ambitious cast, many of whom have fallen prey to the desire for simple, concise explanations of exceedingly complex events. They have cited, for example, climate change, population fluctuations, or even the appearance of a solitary genius as the trigger for agriculture.

But the origins of agriculture don't resemble a law of nature at all. Throughout the world, people responded in particular ways to particular circumstances, and their responses resemble something that is much more familiar to us — a narrative. These accounts have plots, characters, settings — all the elements of a good story. And each story is different, which is why it is so hard to come up with a single explanation for the origins of agriculture.

But the stories have common themes, even if they play out differently. One such theme is population growth. In many parts of the world, hunter-gatherer societies expanded in numbers right before the advent of agriculture. In previous millennia, if wild plants and animals became scarce, foraging peoples could simply move. But by 12,000 years ago or so, that option became less viable in many places, because moving meant coming into conflict with other people living off the land. Finding ways to use the land more intensively must have seemed the easier option.

Another common theme is the end of the last Ice Age. For most of the time modern humans have existed, the planet has been locked in a cold spell. During just two periods has that not been the case. One was when modern humans first appeared in Africa, and the other is right now. For the rest of the time, the earth has been much colder than it is today. In the severest parts of the Ice Age, great glaciers covered much of northern Europe and North America, and large areas of the continents were bitterly cold. Farther from the poles, large swaths of Asia and Africa, including the Middle East, were cooler and drier than they are now.

Then, about 16,000 years ago, the world began to warm. The glaciers retreated to the north; the Middle East became warmer and wetter. Plants and animals that had taken shelter in tiny refugia during the cold began to expand their ranges. But the warming was not uninterrupted. At times the fierce north winds returned. Populations of modern humans must have been subjected to a climatic whiplash that put great stress on their growing populations. Among the Natufians, for instance, a sudden drying of the climate about 11,000 years ago must have caused food supplies to dwindle. They undoubtedly began looking for new ways to feed themselves.

A final theme in the transition to agriculture is what might be called greater social complexity. Around the end of the Ice Age, many societies were organizing themselves in new and more complex ways. The

Natufians began burying their dead in cemeteries with elaborate grave goods. Some of their structures appear to have been not homes but shrines. Tools were engraved with patterns and the likenesses of animals. Similar changes were occurring or soon would occur in Europe, China, and the Americas.

The particular circumstances of each region influenced the development of agriculture there. Most important, each region had unique assemblages of plants and animals that were more or less amenable to domestication. The Middle East was especially blessed in this regard. Of the fifty-six wild grasses with the largest seeds, thirty-two grow in the Middle East, including wheat and barley. No other part of the world has more than a few such plants. The wild ancestors of four of the world's most important domesticated mammals — the goat, sheep, pig, and cow — lived in the Middle East.

This lucky confluence of biological resources had a profound effect on human history. The early appearance of agriculture in the region between the Mediterranean and the Persian Gulf caused the first civilizations to arise there. These civilizations produced the first writing systems, new forms of government, and many of the world's great religions. Human history has been shaped to a considerable degree by the plants and animals that happened to live in the Middle East.

The transition from foraging to farming did not have much effect on the nucleotide sequences in human DNA. Indeed, we now have more or less the same DNA sequences as did our hunter-gatherer ancestors. But the transition to farming did have one major consequence for human DNA: it created much more of it.

Populations of hunter-gatherers are inherently limited by the resources they can gather. Where resources are relatively plentiful, populations can expand. But even in the best areas, hunter-gatherers are never numerous. For example, about 13 million people live in modern Israel, the West Bank, Gaza, and Jordan today. Before the invention of agriculture, the population of the region was probably just a few thousand. The population of the entire world right before the transition to agriculture was probably about 6 million — one one-thousandth of what it is today.

In a population that is stable or growing slowly, which was the situation before the invention of agriculture, women must either limit the number of children they have or watch their children die of starvation before adulthood. As might be expected, modern hunter-gatherers generally choose the former option. Women in many contemporary foraging populations give birth only about once every four years. Because foragers are constantly on the move, young children have to be able to walk before a woman has another that she can carry. Children are breastfed for several years, during which time the mother is unlikely to become pregnant. Abortion (using medicinal plants) and infanticide also must have been ways of controlling family size in the past.

After the development of agriculture, the equation changed. Large families could help with planting and harvesting. Once animals were domesticated, children could be weaned from breast milk to goat or cow milk, and mothers could have more children. If a population became too large, grown children could move elsewhere and start their own farms. Any hunter-gatherers in those areas could be chased away, incorporated into farming economies, or otherwise accommodated.

The resulting growth of population was dramatic. From a worldwide total of 6 million right before the invention of agriculture, the number of humans grew to perhaps 250 million by the first century A.D. — about a fortyfold increase. Ten thousand years ago all humans survived by hunting and gathering. By A.D. 1, the majority of people worldwide relied on domesticated plants and animals.

Estimates of population increases can be used to answer an intriguing question. How many modern humans have ever lived on the earth? Assuming that the population grew at a slow but steady pace from 20,000 some 150,000 years ago to 1 million 65,000 years ago and then to 6 million right before the invention of agriculture, a total of about 7 billion modern humans were born during that 140,000-year period. In other words, between the appearance of modern humans and the invention of agriculture, about as many people lived as the number who are alive right now.

With the transition to agriculture, population growth took off. Between 10,000 years ago and A.D. 1, an additional 26 billion humans were born. Between A.D. 1 and 1750, when the industrial revolution spurred another rise in numbers, 32 billion more humans were born. And since 1750, an additional 16 billion people have existed.

Adding up the numbers, approximately 81 billion modern humans have lived altogether. For every person alive today, about twelve have died. If people really go to heaven after death, then the afterworld is a crowded place.

If the origins of agriculture are a story, with a few sharply drawn characters and themes, then the origins of civilization are an immense, sprawling novel. Hundreds of factors are involved, each affecting the others. A partial list might include trade, craft specialization, record keeping, religion, public ideologies, factionalism, social hierarchies, land ownership, technological developments, ecological diversity, natural resource distribution, warfare, differentiation in wealth, and environmental change. Deriving a single explanation for the origins of civilization from this tangled mass of contributing factors is almost certainly impossible.

But one factor seems especially important: population density. Though exceptions exist, civilizations tend to develop where enough people live to support diversified economic and social systems. The Middle East, the first part of the world to undergo sustained population growth because of agriculture, is also where the first civilizations developed. Especially when farmers began diverting the waters of the Tigris and Euphrates rivers onto the rich soils of the surrounding floodplains, they could produce much more food than they needed to feed themselves and their families. Economic and political systems arose to distribute those food surpluses. Crafts workers produced pottery, hoes, plows, and tools of all kinds. Bureaucrats kept track of the harvests and oversaw the exchange of food for other goods. Priests led religious observances. Kings, chiefs, or head priests exerted whatever control they could over the whole.

By 3500 B.C., a massive temple, taller than anything humans had ever built before, had arisen in the city-state of Uruk, 150 miles upstream from the confluence of the Euphrates and the Tigris. The temple served not only as a religious center but as a center for distributing food surpluses. Systems for tracking these distributions were needed, and by 3400 B.C. the first writing appeared. Clay tablets were inscribed with symbols for such common objects as animals or pots. These pictographs gradually became more abstract, culminating in about 2000 B.C. in the wedge-shaped symbols known as cuneiform. Now scribes began to do more than record commercial transactions. They wrote poems, love stories, hymns to the gods, and descriptions of warfare and the destruction of cities.

By 3000 B.C., what is today southern Iraq supported a recognizable civilization, that of the Sumerians. Other civilizations were also in their nascent stages throughout the Middle East. Upper and Lower Egypt were united under their first pharaoh. Trade networks were developing in the Indus Valley region that would lead, in the third millennium B.C., to the Harappan civilization. Other civilizations would follow in a complicated jumble: those of the Elamites, Akkadians, Babylonians, Assyrians, Canaanites, Israelites, and so on. Outside the Middle East, similar processes were under way in China, Greece, Mesoamerica, and elsewhere. Through the development, spread, and independent invention of agriculture, a new kind of world was being created.

Daily life in these early civilizations was different than it had ever been. In hunter-gatherer societies, only infants and the infirm are spared the task of gathering food. With the advent of civilization, an adult could spend many years rising through a temple bureaucracy, fighting against invading barbarians, inspecting irrigation canals, or doing many tasks other than food production. People no longer all engaged in one activity focused on a common goal. Human activities became centrifugal, fragmented, directed toward many goals.

The relationships among people were similarly altered. Before the rise of civilization, individuals spent their entire lives within the context of bands and tribes. As a result, the predominant form of social connection was kinship. People were defined by their relations to others, and the world was divided into two groups — kin (usually the members of the tribe) and everyone else.

The development of civilization did not necessarily reduce the importance of kinship ties. But people developed ties to other institutions as well: a religion, a city, an occupation, a king. In some cases these new bonds of allegiance overlapped with and reinforced kinship ties. In other cases they didn't.

That a tension would develop between kinship and the state quickly became apparent in the first cities of the Middle East, which included people with many different tribal affiliations. Among the early Egyptians were northern Africans, black Africans from south of the Sahara,

Semitic peoples from the Arabian Peninsula, Hittites from Turkey, and Europeans from even farther north. A state draws strength from the cohesion of its citizens, especially when they are asked to fight for or defend the state. The leaders of the early Middle Eastern states therefore faced a challenge: they had to create an allegiance to the state among people accustomed to thinking only in terms of kin.

One way to do this is by fostering the impression that the state is a tribe. The king is portrayed as the father of the people. The citizens of the state are held to descend from a common ancestor; the Romans, for instance, traced their history to Romulus. Even a state encompassing many peoples can become a sort of tribe in the making. As people within a state intermarry and adopt a common culture, they become kin, or at least possible future kin.

However, the state and the tribe are very different entities. When the interests of the state conflict with those of kinship, the state inevitably opts to preserve or strengthen itself. For example, many of the ancient civilizations in the Middle East were chronically short of labor because of the effects of war, disease, or territorial expansion. To alleviate the shortages, states continually imported new people. These immigrants could not immediately be integrated into an official state kinship system (to the extent that such a system existed). So states had to find ways to gain the allegiance of their people without appealing to kinship. One way was to elevate loyalty to the state above other ties. Every day across the United States schoolchildren recite the pledge quoted at the beginning of this chapter: *I pledge allegiance to the Flag, . . . and to the Republic for which it stands . . .* There's no mention of family, or children, or aged relatives, or ancestors — just one nation, under God, indivisible. It's a small example, but it is emblematic of a state's need to broaden the allegiance of its people.

Many states have tried to redirect kinship ties to other purposes, but none has entirely succeeded. People have always maintained affiliations to groups that more closely resemble bands and tribes than they do states. These groups typically are united by particular symbols or cultural attributes, such as a language or a common homeland. They tend to be broader and more diffuse than kinship groups, in part because the existence of the state forces them to adopt some of the state's attributes. Yet their members almost always claim to have a common biological origin.

These are the groups we know today as ethnic groups. Some sociologists have advocated that "ethnic group" be used instead of "race" to describe people with a common culture, regardless of their relatedness to one another. But ideas of ethnicity and ancestry have inevitably remained entangled. Granted, a person can become a member of an ethnic group without being born into it, say through adoption, marriage, or religious conversion. But ethnic groups still define themselves partly in biological terms, both because of their belief in a shared ancestry and because new members generally are expected to marry within the group.

Ethnicity is as much a matter of perception as of reality. It depends on what is believed to be true, not on what is actually true. As we've seen, human groups are in fact highly fluid. They absorb new members, split into parts, are blurred around the edges, and eventually change into entirely new entities. Most of the ethnic groups of the ancient Middle East no longer exist — the Sumerians, Elamites, Amorites, Kassites, Philistines, and many others. They flared into prominence for a moment in time and then disappeared.

Yet ethnic groups have an incontestable reality in the lives of individuals. They furnish at least part of the social setting in which people act and make decisions. Through their proscriptions on thought and behavior, they lend meaning and context to life.

The path from the invention of agriculture to the origins of ethnicity is a winding one, yet a clear chain of cause and effect links one to the other. The complicated mix of kinship, ethnicity, and nationalism that continues to dominate human affairs arose first in the Middle East. And one group in particular provides a superb demonstration of the role that biology plays in this mix — the Jews.

6 ➤➤

God's People

A Genetic History of the Jews

> Now the Lord said to Abram: Go from your country and
> your kindred and your father's house to the land that I will
> show you. I will make of you a great nation, and I will bless
> you, and make your name great, so that you will be a blessing.
> I will bless those who bless you, and the one who curses
> you I will curse; and in you all the families of the earth shall
> be blessed."
>
> — Genesis 12:1–3

THE JEWS FIRST APPEAR in the historical record as a group of tribes living in the hills around the Dead Sea about 1200 B.C. Hostile invaders from somewhere in the Mediterranean had recently settled along what is today the Gaza Strip. These sea people, who became known as the Philistines, had weapons made of iron, not the more common bronze. Now these fierce warriors were threatening to extend their control into the interior.

To counter the threat from the Philistines, the hill tribes formed a loose confederation. Under the command of several inspired leaders, the confederated tribes successfully repulsed the invaders, eventually confining them to a narrow section of the coast. This successful military campaign set in motion a process of political centralization that led, in about 1000 B.C., to the emergence of a more or less unified state. This is the nation we know today as Israel.

Israel was one of many small states in the area the Romans would later call Palestine (a name derived from the word *Philistine*). These states owed their existence to a temporary quirk of regional politics.

All the usual superpowers — Egypt to the south, Babylonia to the east, and Assyria to the north — were undergoing periods of instability. In the resulting power vacuum, the peoples of Palestine were able to control their own political destiny.

But the respite was brief. In 722 B.C. the Assyrians conquered the northern half of Israel, which had fissioned many years before as a result of internal conflicts. In keeping with Assyrian military practices, the conquerors scattered the northern kingdom's upper classes throughout the Middle East and imported colonists from Babylonia and Syria to take their places. The southern kingdom, which called itself Judah, remained independent until 587 B.C. In that year the armies of Babylonia entered Judah's capital, Jerusalem, and burned it to the ground. Judah's leader, Zedekiah, was forced to watch his children murdered and then had his eyes put out. Like the northerners before them, the elites of Judah were exiled from their homeland, with most of them settling in Babylonia. In time they became known as the people of Judah or, more simply, the Jews.

During the period of social turmoil right before and during the exile, Jewish religious leaders sought to capture in writing the basic tenets of their faith. Drawing on oral accounts passed down through the generations, they described the origins of their people many centuries before. They wrote of a man named Abraham, who moved with his wife, servants, and flocks from Babylonia to the land of Canaan. They described how Abraham's descendants traveled to Egypt during a time of drought and were enslaved by the pharaoh. They recounted the rise of a great leader named Moses, who led the Israelites out of Egypt and back to Canaan, where they conquered the indigenous people and gave rise to the tribes that would later unite to form the nation of Israel.

Few things seem more distant today than the names of the ancient peoples of the Middle East. Yet the Jews have become a powerful force in world politics, religion, and culture. Two partly contradictory forces account for their remarkable longevity. The first has been their steadfast belief in a single, all-powerful God who will protect anyone who obeys the commandments laid down in the Torah. The second has been the creation of a vital ethnic identity that emphasizes Jewish distinctiveness even in the midst of cultures that would seek to destroy that distinctiveness.

Genetically, the Jews are one of the most fascinating peoples on earth. It's not that they differ biologically from everyone else. On the contrary, their genes show that they are a relatively typical, though well-traveled, Mediterranean people. Their DNA is interesting for another reason. It shows how the relationship between the Jews and their God has made itself felt at the most basic biological level.

A good place to begin the genetic history of the Jews is with Aaron's Y chromosome. In the book of Exodus, God decrees that Moses' brother Aaron and all of his male descendants shall be the high priests of the Israelites. Even today, men who count themselves among the direct male descendants of Aaron have special responsibilities in many synagogues, such as leading certain blessings. Within Judaism as a whole, these men are known as *kohanim,* the Hebrew word for priests. Many have the last name Cohen, Cohn, Kahn, or a similar derivative of the word *kohan.*

Since men pass their Y chromosomes on to their sons, all of Aaron's sons would have had his Y chromosome, which they in turn would have passed on to their sons, and so on down the generations to the kohanim of today. Along the way, mutations would occur in the separate lineages derived from Aaron's Y chromosome, making the nucleotide sequences of Aaron's male descendants somewhat different from one another today. But the original haplotype should still be visible, like a figure behind a translucent screen.

A few years ago a team of geneticists from Haifa Technion, University College in London, and the University of Arizona set out to find Aaron's Y chromosome. Using cells swabbed from the cheeks of about two hundred Jewish males from Israel, North America, and England, they looked for specific genetic markers along each man's Y. They found that Jews who did not identify themselves as kohanim had a broad assortment of Y-chromosome markers, no one of which was especially frequent. But of the kohanim, about 50 percent had a particular set of markers, indicating that all of their Y chromosomes descended from a common ancestor. The researchers called this genetic pattern the Cohen Modal Haplotype (modal in this case meaning most common).

The more recent mutations among men carrying this haplotype also allowed the researchers to calculate when it originated, just as the age of mitochondrial Eve has been calculated from the differences in our mitochondrial DNA. According to the geneticists' calculations, the man who carried the original chromosome lived about 106 generations ago. Within the margin of error inherent in the calculation, this easily falls within the time frame when Aaron may have lived.

Other explanations could account for the overrepresentation of the Cohen Modal Haplotype among the kohanim. At some point early in the religion of the Israelites, a single man, or a group of men descended from a single male ancestor, could have established themselves as priests. So long as rival claimants to the priesthood were smaller in number or had fewer sons, a particular Y chromosome could have become more common among the kohanim. Yet these explanations seem even more complicated than the one in the Bible. Given the specificity of the story, the most historically parsimonious explanation is that Aaron really existed.

Other measures of Jewish genetics do not accord so neatly with biblical accounts. The Hebrew Bible says that the forefathers of the Jews were the twelve sons of Jacob, whose descendants settled the land of Canaan after the exodus from Egypt. Furthermore, Jewish religious laws call for strict endogamy — that is, Jews are supposed to marry other Jews. As Deuteronomy 7:3 says, "Do not intermarry with them [non-Israelites], giving your daughters to their sons or taking their daughters for your sons, for that would turn away your children from following me, to serve other gods. Then the anger of the Lord would be kindled against you, and he would destroy you quickly." If this edict had been scrupulously observed since the time of Jacob — and if there had been no converts to Judaism — then all Jewish males would have Jacob's Y chromosome.

Jewish history has obviously been much more complicated than that. In a recent study, a team of twelve researchers from the United States, Israel, England, and South Africa found that Jewish males have a wide variety of Y-chromosome haplotypes — far too many to have come from a single man living anytime in the past few thousand years. Rather, most of the Y chromosomes found in Jewish males are the same as those found among many other men from the Middle East — Palestinians, Syrians, Lebanese, Saudi Arabians, and so on. The Jews and their neighbors all emerged from the same diverse pool of Middle Eastern people.

Studies of mitochondrial DNA yield similar results, with one slight exception. The mitochondrial DNA sequences of Jewish females are even more diverse than the Y chromosomes of males, suggesting that non-Jewish women converted or married into the faith even more often than did men.

These results are hardly surprising. Following the defeat of Babylonia by the Persians, the Jews in exile were allowed to return to their homeland and rebuild their state. Over the next few centuries, many Middle Eastern people converted to Judaism, not only in Israel but in Jewish communities scattered throughout the Middle East. By A.D. 1 the total Jewish population in the Middle East probably exceeded 5 million.

By then the rapid growth of Judaism had brought it into direct conflict with the newest Middle Eastern superpower — Rome. After a series of Jewish rebellions against Roman rule, the legionnaires moved in. Following a protracted siege, the Roman armies fought their way into Jerusalem in A.D. 70. They massacred the city's inhabitants, burned the Jewish temple, and sold the survivors into slavery. A much weakened Jewish state persisted for several more decades, but in A.D. 135 the Romans put an end even to that.

From this point on, the Jews practiced their religion only as part of the Diaspora — the far-flung network of Jewish communities scattered from Spain to China to southern Africa. Yet they kept alive the memory of their homeland. They practiced their religion in synagogues that served as surrogates for the temple in Jerusalem. They looked forward to the day when the state of Israel would rise again in the Middle East.

When that day finally arrived eighteen centuries later, the Jews were a different people genetically than they had been when they left.

Ask a geneticist how best to learn about the genetic history of the Jews, and you will almost always get the same reply: "You should go talk with Batsheva." A professor of human genetics at the Tel Aviv University School of Medicine, Batsheva Bonné-Tamir is head of the National Laboratory for the Genetics of Israeli Populations. For four decades, almost every major paper on the genetics of Jewish populations has listed her name among the authors.

I met Bonné-Tamir at her home in a small town near Tel Aviv. She's tall, has bright red hair, and speaks English with just the slightest trace of an Israeli accent. We sat in her backyard surrounded by kumquats, oranges, and hibiscus. "Israel is a Garden of Eden for studying genetic diversity," she told me, sweeping her arm toward the fields beyond her

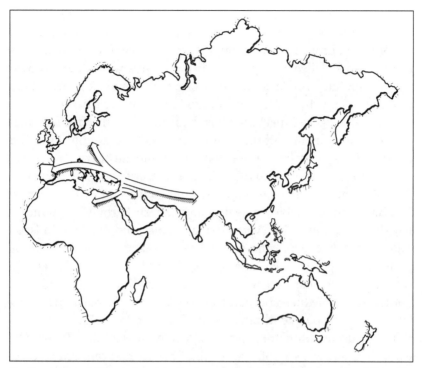

Starting in the eighth century B.C., the Jews scattered from their homeland through-out the Middle East and into Africa and Europe. Later migrations took Jews into northern Europe and parts of Asia.

fence. "We're a small country, but our people come from everywhere. That's why geneticists are so eager to work here."

Bonné-Tamir grew up in Jerusalem, the daughter of academic parents who had emigrated from Germany to Palestine. As a child she was surrounded by Jews, but Jews who were strikingly different from one another: a maid from Yemen, Mediterranean Jews from Morocco and Algeria, Jews from northern Europe, even Jews from Africa. "I was interested in inheritance versus the environment," she told me. "I wanted to know which was more important."

If the Jews originated from a single group of people, how could Jewish groups have developed such distinct physical appearances? One admittedly unlikely possibility is that the Jews who settled outside the Middle East developed new traits in response to the environments where they lived. For example, maybe those in northern Europe developed lighter skin in much the same way that modern humans moving north from Africa did many thousands of years ago. This environmental explanation seems unlikely, however. Jewish populations have lived in northern Europe for only about fifty generations. Only if many Jews in these areas failed to reproduce because they were underexposed to the sun, and thus deprived of vitamin D, could skin color have changed so quickly, and no evidence for any such trend exists.

The more likely explanation involves the same mechanisms that were operating in ancient times: admixture and conversion. As Jewish populations spread from the Middle East into northern Africa and Europe, most Jews married other Jews, as prescribed by their religion. But over time, marriage to non-Jews and conversion of non-Jews to Judaism were inevitable. In central and eastern Europe, for example, Jewish populations mixed with Germans, Poles, and Russians. In Israel today the Ashkenazim, the Jews whose ancestors came from those parts of Europe, are taller and lighter-skinned than other Israelis. In contrast, the Sephardic Jews, whose ancestors lived in the countries bordering the Mediterranean, tend to be darker and smaller, reflecting the traits of the people whom they lived with and married.

The issue of admixture has always interested Bonné-Tamir. She works in a medical school, and much of her current research involves genes that cause specific diseases. But admixture is an important consideration in medicine, too. Jews returning to Israel tend to have somewhat different sets of medical problems. To understand these differences, doctors needed to know more about the genetic history of the various Jewish populations.

Early in Bonné-Tamir's career, Israeli researchers sought to compare the genes of different populations using what are known as classical genetic markers — biological characteristics, such as blood groups and immunoglobulins, that differ in frequency from population to population. "When the planes carrying immigrants arrived, physicians would be waiting at the airport to get blood samples," she says. But the results from tests of these markers were ambiguous. Some indicated that various Jewish groups were very similar and probably had common origins. Others suggested that different Jewish populations couldn't possibly be related — lending support to such ideas as the claim that the Ashkenazim are descended not from Middle Eastern Jews but from Turkish and Slavic peoples who converted to Judaism during the Middle Ages.

Only in the past few years have biological techniques become sophisticated enough to arrive at firmer conclusions. Genetic research has now demonstrated, for example, that the Ashkenazim are indeed descended from Jews who migrated to Europe from the Middle East. Separate studies of their Y chromosomes and mitochondrial DNA have revealed characteristically Middle Eastern haplotypes. The Ashkenazim and their European neighbors intermarried enough to introduce some mitochondrial DNA and Y chromosomes from elsewhere and to influence physical appearance. But estimates of admixture are lower than expected. Based on data gathered from the Y chromosome, fewer than one in one hundred Jewish women had children with non-Jewish men in each generation.

The same genetic techniques are revealing the flip side of admixture. As Jews have scattered around the world, they have introduced their genes into populations not normally considered Jewish. The best-known example involves the Lemba, an ethnic group of black Africans, numbering perhaps 50,000, scattered throughout southern Africa, with a particular concentration south of the Limpopo River. Lemba tradition says that their ancestors came by boat to southern Africa from a northern town called Sena — a name similar to town names in Israel, Egypt, Ethiopia, and Yemen. Though the Lemba generally resemble their Bantu-speaking neighbors, they say that their ancestors were Jewish, and they practice a number of Jewish-sounding rites, such as particular food taboos and circumcision.

In the past, most historians dismissed the Lemba's claims to Jewish ancestry. Perhaps the ancestors of the Lemba had read parts of the Hebrew Bible and adopted some of its practices. Or maybe they absorbed Jewish traditions from traders on the east coast of Africa. A surprising number of groups around the world claim to be descended from Jewish ancestors, though the historical connections between most of these groups and the ancient Hebrews appear tenuous at best.

In 1997 a team of geneticists decided to take the Lemba's claims more seriously. They obtained blood samples from a number of Lemba males and analyzed their Y chromosomes. Though about one-third of the Y's were clearly those of Bantu speakers, two-thirds were of Middle Eastern origin. What's more, the majority of the Y-chromosome haplotypes among the oldest and most influential of the Lemba's clans were not just generic Middle Eastern chromosomes. They were the Cohen Modal Haplotype — the marker chromosome for the Jewish priesthood.

The genetic study of the Lemba still has not determined the antiquity of their Middle Eastern ancestry. Perhaps the ancestors of the Lemba mated with Jewish traders over the past millennium and the children born of those unions stayed in Africa. Or maybe the Lemba really do descend in part from a group of people who migrated from the Middle East during the Diaspora. So far, study of their mitochondrial DNA has revealed no evidence of female ancestors from the Middle East, so their female ancestors must have been overwhelmingly African. But at this point their Jewish ancestry on the male side seems assured.

This connection between a group of black Africans and the Jews may seem surprising at first. Yet more and more of these kinds of genetic links will be found as more is learned about our genetic heritage. The forces of genetic mixing are so powerful that everyone in the world has Jewish ancestors, though the amount of DNA from those ancestors in a given individual may be small. In fact, everyone on earth is by now a descendant of Abraham, Moses, and Aaron — if indeed they existed.

All this talk about the genetic diversity of the Jews raises an obvious question. The religious admonition to marry within the faith would seem to be a force for genetic homogeneity, not heterogeneity. As people marry within the same group over many generations, they end up marrying distant relations, which has the effect of reducing the genetic variety within a population. And isn't such intermarriage ultimately a bad thing? For example, don't Jews suffer from a larger number of genetic diseases than other people because they generally marry each other?

When I asked Bonné-Tamir these questions, she had a simple suggestion. "Let's go see the Samaritans." So we headed east in my rental car. Twenty minutes from the outskirts of Tel Aviv we were at the checkpoint leading into the West Bank, and in another twenty minutes we reached the outskirts of Nablus. We turned off the main road at the edge of town and started up a long series of switchbacks. High above us a cluster of white buildings fringed the ridgeline — the Samaritan homeland on Mount Gerizim.

Bonné-Tamir has been working with the Samaritans since the early 1960s, when she was a graduate student at the University of Chicago. To earn a master's degree in physical anthropology, she had to write a thesis about an indigenous group. Most of her classmates were studying Native American tribes, but Bonné-Tamir wanted to work on a topic closer to her own home. At the Hillel Library on campus, she came across a book about the Samaritans. She wrote to them and found a young man who was willing to correspond. Her first published paper, a 1963 adaptation of her master's thesis, was entitled "The Samaritans: A Demographic Study."

The Samaritans trace their history to the northern kingdom of Israel. Their version of the Torah is still written in an ancient Semitic script. Many of their practices predate the fall of the southern kingdom in 587 B.C. But when the people of Judah began to return to Israel after the Babylonian exile, they never quite accepted the Samaritans as proper Jews and rejected proposals to reunite. Since that time the two traditions have evolved in parallel, with striking similarities but obvious differences as well.

One thing the Jews and Samaritans share is a history of persecution. Samaritan communities expanded under the Romans (the Good Samaritan of the New Testament belonged to one such community) but then were severely oppressed. Their communities in Syria and Egypt went extinct, and by the end of the nineteenth century the total number of Samaritans had declined to only about 150.

In the twentieth century the community began to grow again, and it now numbers more than 600. Half live in the town of Holon outside Tel Aviv. The other half live on Mount Gerizim. I visited during the week of Passover, and all the Samaritans were there, the Holon contingent staying in the summer homes they've built on the mountain.

The Samaritans are the most inbred population known anywhere in the world. They have married within the religion for more than 2,000 years. In fact, they now marry largely within five specific male lineages, two in Mount Gerizim, two in Holon, and one split across the two communities. In more than 80 percent of marriages, the partners are either first or second cousins.

If high levels of intermarriage were a serious health risk, one would expect the Samaritans to be genetic basket cases. That's obviously not so. During the afternoon I spent on Mount Gerizim, the Samaritans struck me as a hearty, proud, and generous people. "We are touching paradise here on this mountain," said Benyamim Tsedaka, the coeditor of *A.B. — The Samaritan News* and a prominent spokesman for the community. "We don't have to wait until the next life." Over innumerable cups of tea in living rooms up and down the main street, the only unusual thing I noticed about the Samaritans was that many relatives bear an uncanny resemblance to one another. An aunt and her niece will look like twins, despite being separated in age by twenty years. Cousins look as closely related as brothers.

The Samaritans do have a few genetic problems. Some of their members are congenitally deaf because they have inherited two copies of a defective gene on chromosome 11. A movement disorder caused by a different gene affects a few more of their members, forcing them to walk with a cane. The Samaritans have now identified the carriers of these genes, and they take steps to ensure that carriers do not marry. "When young people want to get married, nobody thinks about the genetics," said Tsedaka. "We have learned to be more careful."

Many Western countries have strong legal and religious sanctions against marriages between close relatives, and many people believe that such marriages inevitably lead to genetic disaster. Even the word "inbred" has a pejorative connotation (geneticists prefer "consanguineous" — a wonderful word derived from the Latin word for blood). But this Western attitude is the exception; in many Asian and African countries, marriages between cousins are common.

If such marriages wreaked genetic havoc, the world would be full of genetically enfeebled populations. Instead, careful analyses have shown that consanguineous marriages affect the health of only a small percentage of offspring. Studies document a slight rise in miscarriages and in health problems that can cause an infant to die. But the women in consanguineous marriages tend to have more children than other women, usually because they marry earlier. As a result, overall rates of fertility are quite similar between groups that marry in and those that marry out.

The critical factor for the genetic health of a population is not the number of consanguineous marriages, Bonné-Tamir explained to me. Groups that have married in for long periods tend not to have any more genetic problems than more mixed populations. But the kinds of problems they have are related to the group's particular genetic history. The Samaritans, for example, went through a severe population bottleneck at the end of the nineteenth century, when they came close to dying out altogether. The people who survived had very few genetic problems, and today their incidence of genetic diseases is not higher than for other groups. One hypothesis even says that severe bottlenecks weed out defective genes by exposing them to especially severe selective pressures.

Similarly, the genetic diseases associated with Jewish populations result from the history of those populations, not from the effects of close marriages. About 10 million Ashkenazim are alive today — they make up more than three-quarters of the 13 to 14 million Jews in the world. Yet the great majority of their DNA probably comes from a few thousand Jews who settled in central Europe in the Middle Ages. The Ashkenazim tended to have large families, so their numbers grew dramatically. As a result, a relative handful of harmful genes present in the founder groups spread widely.

Every population has its own history and thus its own susceptibility to a particular set of genetic diseases. The Finns, who also seem to be descended from a relatively small number of founders, have a different set of defective genes. Europeans in general have an elevated frequency of a mutated gene that causes cystic fibrosis when inherited from both parents. The story is the same for other groups, from sickle cell disease in Africans to various cancers in Asians. Because Jews are more identifiable as a group, their genetic diseases are more visible. But Jews overall do not suffer from more genetic diseases than do the members of any other group.

I suspect that as you've been reading this chapter, you've occasionally felt a strong sense of foreboding. How much do we really want to know about the genetics of groups — aren't there great dangers associated with this work? During World War II, all Jews in Germany had to wear a yellow Star of David emblazoned with the word *Jude*. Anyone who had one Jewish parent was considered a Jew because, as a German official put it, "among half-Jews the Jewish genes are notoriously dominant." Having already been identified, German Jews were easy to gather into concentration camps when the war started. By the end of the war the Holocaust had reduced the worldwide Jewish population from a prewar high of 16.6 million to 11 million.

At the individual level, people have to decide how much they want to know about their own ancestry. For example, say that a man who considers himself a member of the kohanim is obsessed with a desire to know if he has Aaron's Y chromosome. But the odds of the test for the Cohen Modal Haplotype coming up positive are about fifty-fifty, so if he has a lot invested in getting a positive result, he would almost certainly be better off not being tested. As with paternity, what you don't know usually can't hurt you.

Then again, for some people the information provided by a test of genetic ancestry can provide a powerful sense of connectedness. Even if a test fails to demonstrate a particular ancestry — such as descent from Aaron — it will inevitably demonstrate ancestry from someone else. Furthermore, future genetic tests will be able to reveal not just one line of descent but many. Today tests are available only for mitochondrial DNA and the Y chromosome, but these tests detect only single lineages extending back into the past. Soon tests of genetic markers on the chromosomes will reveal the full diversity of our ancestry. Geneticists will be able to show, for example, that a person received a piece of chromosome 21 from Jewish ancestors, a piece of chromosome 3 from African ancestors, and so on. At that point genetic tests may become distinctly less compelling, because they will simply demonstrate the obvious: that everyone is related to everyone else.

Groups, too, will have to decide just how much they want to know about their genetic past. After all, the genetic tests of the Lemba could easily have shown that they had no biological connection to Jewish ancestors. Such results probably would not change the Lemba's ideas about themselves, because genetic results that don't line up with history can usually be accounted for. Still, to the extent that groups have control over the testing of their members, they will have to prepare themselves for what genetic tests might reveal.

The fear expressed most often about genetic testing relates to both individuals and groups. It may seem that such tests could unequivocally identify individuals as members of particular groups, functioning as a sort of genetic yellow star. But genetics research may turn out to have exactly the opposite effect. For example, no single genetic marker can indisputably prove that a person is Jewish. Even the Cohen Modal Haplotype, Aaron's Y chromosome, is found in all other Middle Eastern populations, both because of intermarriage and because that haplotype existed before the Jews took shape as a people. If every male with the Cohen Modal Haplotype instantly converted to Judaism, the politics of the Middle East would be turned upside down. Similarly, because Jews have mixed with non-Jews throughout history, genetic markers that arose in Jewish populations are now scattered around the world. If a particular genetic marker were enough to make a person Jewish, the recounted Jewish population would grow manyfold.

These observations about individuals and groups need to be treated carefully. Statistical associations can be made between collections of genetic markers and particular groups. For example, geneticists might someday be able to say that if a person has these ten markers and doesn't have those ten, he or she has a certain probability of being a member of a particular culturally defined group. But the associations will always be statistical because very few people will have exactly the same set of markers — and because the boundaries of the group will always be poorly defined. Our histories are too interconnected to draw distinct lines between collections of people.

If anyone has a right to be worried about the potential misuses of genetics research, the Jews do. But in conversations with Jewish friends and Israelis I've received quite another impression. People are proud of their history and want to know more about it. If that knowledge has medical benefits, all the better. "I've been doing genetic and medical research in Israel since the 1960s, and I've never been accused of scientific racism," says Bonné-Tamir. "I've had excellent rapport with every group I've studied."

The genetic history of the Jews reemphasizes a conclusion that comes up repeatedly in the study of human groups. Our DNA forms a genetic substrate that everyone shares equally. Only human cultures can draw sharp distinctions among people. "Being Jewish doesn't have anything to do with your genes," a friend once told me. "It has to do with who you are."

III

Asia and Australia

7

The Great Migration

To Asia and Beyond

> There is a Chinese saying,
> "Either the East Wind prevails over the West Wind
> or the West Wind prevails over the East Wind."
>
> — Mao Zedong, Moscow Meeting of Communist
> and Workers' Parties, November 18, 1957

"THIS ONE WAS a woman," says Zejing Tan, handing me the small, white skull. "She was young — see, these bones hadn't fused yet." I turn the skull upside down to peer into its interior. The bones that underlie the face are thin and porous, as delicate as feathers.

"European?" I ask.

"No, Mongoloid," Tan replies. She plucks a different skull from one of the blue plastic boxes stacked against one wall. "This one's Caucasian. See how the base of the nose sticks out here," and she points to a saddle-shaped indentation between the eyes.

"It's like your nose," says geneticist Li Jin, who's been translating for me this morning with Tan and with Yongqin Xu, the head of the Anthropology Department here at the Museum of Natural History in Shanghai. "Asian faces are flatter," says Jin, and he draws his hand lightly across his forehead, as if wiping his brow.

I turn the skull right side up again, so that its vacant eye sockets are staring at me. I'm standing amid a community of the dead. Three thousand years ago the people whose bones are in this room shared meals, exchanged jokes, made love. Now their remains are together again, side by side.

These bones are a long way from home. They come from the Tarim Basin, a vast inland desert 2,500 miles northwest of Shanghai. Now part of Xinjiang, the westernmost region of China, the Tarim Basin is as close to the center of Asia as you can get. It rests between snow-covered mountains on three sides — the Kunlun and Tibetan plateau to the south, the Pamirs to the west, and the Altai to the north. Much of the basin consists of the Taklimakan Desert (the popular etymology for the name is "you go in but you don't come out"). On the northern and southern fringes of the desert, oases have formed around the streams that spill from the mountains and lose themselves in the sandy wastelands. The two main branches of the Silk Road, which was established in the second century B.C. as a trade route between the Middle East and China, passed through these oases. This was the most dreaded part of the journey. Traders en route between the Middle East and China feared the isolation, the shifting sand dunes, the brigands who would sweep down from the hills.

Early in the twentieth century, archaeologists exploring the Tarim Basin began to come across ancient cemeteries half buried in the desert sands. Inside many graves the bodies had decayed to bones, like those in the Shanghai museum. But in other graves the dry air and bitter cold of the desert had wrought a miraculous transformation: the bodies had shriveled and dried but were otherwise well preserved. They were dressed in colorful tartans, felt hats, and leather coats. Their skin was so lifelike that the tattoos on their arms were clear and sharp. They almost looked asleep rather than dead. Yet the Bronze Age artifacts at their sides showed they had been lying in their graves for more than 3,000 years.

Even more surprising, many of the mummies had red or blond hair, deep-set eyes, and sharp, protruding noses. Here, on the western edge of China, more than a thousand years before the Silk Road existed, people with European features were living in long-established communities.

The discovery of the mummies has been controversial in China. The largest of the ethnic groups in Xinjiang today is the Uighur, who have both Asian and European ancestors. In their efforts to achieve greater autonomy from China, the Uighur have seized upon the mummies as proof that they are descended from non-Chinese ancestors. An artist's reconstruction of one of the mummies, known as the Beauty of Krorän, has become a popular Uighur icon.

As is usually the case when archaeology is enlisted to support nationalist claims, the reality is much more complicated. People with Asian features also lived in Xinjiang, though not as early as the people with European features. The Uighur are descended largely from people who migrated into the Tarim Basin during the ninth century A.D., though they undoubtedly intermarried with the earlier inhabitants. The skulls in the Shanghai museum, which come from a graveyard near the town of Hami and date from the first and second millennia B.C., exhibit a wide range of features. "We have definitely observed a mixture of Caucasian and Mongoloid individuals," says Xu. "We've also found intermediate types, which makes it hard to draw a clear distinction between groups. Some of the bones are closer to eastern European people, and some are closer to central Asian people. There have been all kinds of people in western China."

This mixing of people, the points of contact between East and West, is what has drawn geneticist Li Jin here. His research team in China has been extracting mitochondrial DNA from the teeth of the Tarim mummies. So far the mitochondria from skulls with European features have had haplotypes found largely in Europe, while those from Asian skulls have had largely Asian haplotypes. But in some cases the skulls and haplotypes don't match up so neatly.

Doing research at the Shanghai museum has been something of a homecoming for Jin. For most of each year he does genetics research in the United States, initially at the University of Texas in Houston and more recently at the University of Cincinnati. But he grew up in Shanghai, just a block away from this museum. He came to this building often as a child, standing in the dark, dusty rooms for hours and staring at the specimens of plants and animals gathered from all over China.

In 1966, when Jin was five, his mother and father, who were both teachers, were sent away from Shanghai because of the Cultural Revolution. An only child, Jin was left behind to raise himself. His formal schooling was spotty. He spent many hours building things: model airplanes, ham radios, model ships. Occasionally his neighbors in the apartment building where he lived looked in on him. Mostly he was alone.

When the Cultural Revolution ended in 1977 and his parents returned, Jin began to attend school more regularly. He quickly proved to be a superb student, especially in mathematics. He earned admission to Fudan University in Shanghai, one of the best colleges in China. That was where he acquired the name by which he's known in the United States today. His English teacher at Fudan asked her students to choose an English name to use in class. Jin looked through a dictionary of names and picked Felix because the name means happy and contented (which is actually a pretty good self-assessment). "Unfortunately," he says, "the dictionary didn't say anything about a certain cartoon character." Too late — he's been Felix ever since.

Jin's parents wanted him to go to medical school, but he wanted to become a mathematician. They compromised. He majored in genetics, the most mathematical of the biological sciences.

Jin was an undergraduate in the 1980s. When the students at Fudan began to call for greater government openness, he was an eager participant — "a troublemaker," he says. In 1988 — a year before the crackdown in Tiananmen Square — he moved to the University of Texas to enter a Ph.D. program in genetics. When he finished, he, like many other Chinese graduate students, realized that it was not a good time to return to China. So he did a postdoc at Stanford University, then landed a faculty position at the University of Texas. He received several large federal grants to study the genetics of arthritis and heart disease. His lab grew quickly — he was a rising star in biomedical research.

But he did not want to lose touch with China. He had married a Chinese woman, and his children were growing up without knowing their parents' homeland. Moreover, the government's attitudes were changing; Chinese officials realized that they needed to attract scientific talent back to the country. Jin negotiated an arrangement in which he would work for three months of each year at Fudan University and the other nine months at Texas. He's been especially active in building up the genetic sequencing centers with which China hopes to join the biomedical revolution. "It's a way for me to make a contribution to the country where I was born," he says.

Jin's research in both countries focuses mostly on biomedical problems. At the same time, he has done more than anyone else to piece together the genetic history of Asia. He has traced the movements of modern humans out of Africa to Australia. He has reconstructed how people moved north into the mainland of China and how they fractured into the many different ethnic groups that live in Asia today. He has shown how Asians continually came into contact with westerners as they spread out to occupy half the world.

Like the study of the Tarim mummies, Jin's research has been controversial. His genetic findings often conflict with beliefs the Chinese have about their own past. Jin says that he goes where the science leads him and lets others worry about the implications. "China's population is very large, and its people are very heterogeneous," he says. "This is a precious resource, to understand how people migrated to Asia and how they have mixed since then."

Someone standing in the far northeastern corner of Djibouti, the small African country at the southern end of the Red Sea, can smell the continent of Asia. The coast of Yemen on the Arabian Peninsula, seventeen miles away across the Strait of Bab el Mandeb, is just out of sight over the horizon. But when the easterlies blow in the fall and winter, the dust of the peninsula streams in great clouds across the strait, and flocks of birds making their way over the waves bear the promise of nearby land.

Modern humans were living on this shoreline right before they began their expansion into the rest of the world. Paleoanthropologists have found the remains of an ancient raw bar on the beach — great piles of shells left just where the ancient diners threw them. No human fossils have been found with these shells, but the piles are obviously the work of humans, and modern human fossils elsewhere in eastern Africa date to about the same period. As those people walked along the western shore of the Red Sea, they must have looked across the water and wondered what lay on the other side.

We may never know how modern humans got from the Horn of Africa to the Arabian Peninsula. Even when the sea level was at its lowest, the strait at the mouth of the Red Sea was at least several miles wide. If they built rafts to cross, we will never find evidence of it, because the wood and lashings would long since have decayed into dust. Of course, they could have walked around the Red Sea — north along the coast and across the Sinai. But between 80,000 and 65,000 years ago, which is the most likely period for the migration, Neandertals still occupied the Middle East, so a short ride on rafts across the Strait of Bab el Mandeb may have been the easier option.

A waterborne migration seems likely for another reason. Once modern humans reached the Arabian Peninsula they kept on going — around the Persian Gulf, along the shorelines of Iran and Pakistan, south along the Indian coast, north again to the mouths of the Ganges, and finally into southeastern Asia. By about 65,000 years ago, according to the current evidence, modern humans had reached Australia.

Along the way they undoubtedly encountered archaic humans. *Homo erectus* had been living in eastern and southeastern Asia for hundreds of thousands of years before the arrival of modern humans. But archaic humans never made it to Australia. Reaching the continent, even during periods of lower sea level, requires an ocean passage of more than sixty miles, so land wouldn't have been visible to anyone setting sail. But if modern humans had been skipping their way along the coast by boat, they may have been prepared to take the longer journey needed to reach Australia from the Asian southeastern corner of the continent.

As with migrations on land, we tend to think of these long journeys as taking place during a single lifetime, as if a small band of modern humans set out from Africa and paddled to eastern Asia. But this expansion undoubtedly took many generations. Probably more than one group came out of Africa, and different groups may have overlapped and mixed along the way. The distance traveled by modern humans during a single lifetime was probably not great. Yet even if they extended their range by just fifty miles in each generation, they could have covered the distance from Africa to eastern Asia in a few thousand years.

Then again, they might have traveled more quickly. If they had tarried along the way, they would have built substantial settlements. But the archaeological record is blank. For all the evidence we have of the initial passage of modern humans across southern Asia, they might as well have flown.

Actually, archaeological evidence of their journey may exist, but we can't obtain it. If modern humans stayed close to shore, most of their camps are probably under water today. During the time when they were making their way eastward, the sea level was considerably lower than it is now. When the level rose, 10,000 or so years ago, the water would have inundated any campsites close to the shore. The stone tools of modern humans are probably just offshore in the warm, murky waters of the Indian Ocean. But until the earth enters another Ice Age and sea levels drop, those tools will be very difficult to retrieve.

Even if modern humans left no archaeological record of their passage, they left a genetic record. Sometime more than 60,000 years ago a mutation occurred in the mitochondrial DNA of a woman living in eastern Africa (specifically, a cytosine mutated to a thymine at position 10,400 of her mitochondrial DNA). Through the workings of genetic chance, the haplogroup defined by this mutation, known as haplogroup M, became common among the populations living in the area. A substantial portion of the people who made their way out of Africa onto the Arabian Peninsula carried mitochondrial DNA from haplogroup M (most of the others had mitochondrial DNA in a related haplogroup labeled N). Today haplogroup M is found in the southern part of Arabia and in India and is widespread in Asia. But it is rare in the Middle East, which is another argument for an early migration route from the Horn of Africa rather than through the Sinai.

Once modern humans reached Australia, they spread quickly across the landscape. Starting about 65,000 years ago, a number of large Australian animals became extinct — presumably hunted down by the continent's new occupants. The vegetation of Australia also underwent major shifts around this time, although the climate stayed about the same. Humans seem to have been setting fire to the grasslands to aid in the hunt.

A skeleton dated to 62,000 years ago has been found near Lake Mungo in southeastern Australia. The remains are clearly modern, with slender limb bones and a high, domed skull. And at a remarkable site in northern Australia, the initial colonists of the continent seem to have hollowed out an array of indentations on the face of a rock — perhaps the earliest instance yet known of symbolic thinking.

The aboriginal people of Australia have long been of special interest to physical anthropologists. Some have traits that are unusual elsewhere in the world, such as a heavier build and more prominent brow ridges. Anthropologists have often speculated that the aborigines might represent an older population of humans, an anthropological relict of an earlier age.

Tests of their DNA have put an end to these speculations: the aborigines have more or less the same genetic variants as all other non-Africans. Maybe their striking physical characteristics reflect the comparative isolation of their continental home. (Although the genetic evidence also seems to point to a relatively recent migration of people from India to Australia, perhaps as recently as 4,500 years ago, when new kinds of stone tools and the dogs known as dingos first appeared on the continent.) But complete isolation is not necessary for the development of characteristic features. It happened many times — in

Asia and elsewhere — as groups of modern humans moved away from their African homeland.

By about 40,000 years ago, modern humans were living in the interior of southeastern Asia. Fossils dating almost to that age have been found in Malaysia, Thailand, Vietnam, and the Philippines. By perhaps 26,000 years ago, modern humans were living as far north as modern-day Beijing.

If a western paleoanthropologist were asked where those first modern southeastern and eastern Asians came from, he or she would probably say, from the populations living farther south, in Australia and New Guinea. But a Chinese paleoanthropologist would probably give a very different answer.

For many years, most paleoanthropologists in China have adhered to a strictly multiregional, or even Sinocentric, view of human origins. They believe that modern Chinese evolved from earlier forms of human beings who lived in eastern Asia. According to this view, modern Chinese descend from "Peking man," a *Homo erectus* fossil found near Beijing that is about 400,000 years old. Almost all Chinese paleoanthropologists firmly reject the idea that modern Chinese descend from people who came out of Africa within the past 100,000 years.

In his efforts to put the genetic history of Asia on a solid footing, Li Jin first had to address the idea that at least some of the Chinese people's ancestors were Asian *Homo erectus*. The notion is not implausible. As modern humans traveled east from Africa, they would have had plenty of opportunities to mate with any humans already living in Asia. In that case, maybe it was a mixture of modern and archaic genes that led to the distinctive appearance of most eastern and southeastern Asians today.

Jin approached the problem the same way geneticists had in Europe. If moderns interbred with archaics in Asia, then traces of the archaics' genes ought to exist in the chromosomes of Asians today. In Europe, geneticists had looked for Neandertal DNA in mitochondria. By the time Jin began his work, he had an advantage. Laboratory techniques had improved to the point that he could look at the much longer Y chromosome. If he could find a single Y that differed from the Y's of other males, it could very well have come from a *Homo erectus*.

He and his students fanned out across China and collected cells from more than 10,000 males. (Though informed consent is not required in China, Jin had each man read and sign a consent form, applying the same ethical standards as he would in the United States.) In all 10,000 Y chromosomes, not a single unusual one was found. "We looked," says Jin. "It's just not there. Modern humans originated in Africa."

The genetic evidence also indicates the direction of movement of modern humans in Asia. Jin has studied markers on mitochondrial DNA and on the Y and other chromosomes — particularly a stretch of chromosome 21 in which recombination has not muddled the genetic signal. All suggest a gradual population movement from south to north, with a thick overlay of subsequent migrations and intermarriage. "There are two kinds of migrations," says Jin. "One is colonization, where people are going to an empty space. The other is gene flow, where an incoming group mixes with people who are already there. To distinguish between the two, you have to look at a large number of populations from many different places. If you see a pattern that is consistent across those populations, then you can have more confidence in the geographical origins of those groups."

In Asia, says Jin, levels of genetic variation are greater in the south, which would be expected if modern humans arrived there first. Furthermore, the haplotypes in the north are a subset of those in the south. Just as happened when modern humans left Africa, the people moving north took just a portion of the south's genetic variety with them. This genetic differentiation may even account for some of the physical differences between northerners and southerners today. The northern Chinese tend to be paler and taller, with smaller eyes and a more pronounced epicanthic fold. The southern Chinese are darker and broader, with features more like the people of southeastern Asia.

To a northern Chinese, the idea that he or she descends from southerners is about as welcome as the news that all Chinese are descended from Africans. In China the north is *zhongguo*, the central kingdom, the source of all civilization and culture. For example, most Chinese — southerners as well as northerners — will tell you that they are descended from the Yellow Emperor — Huang-Ti, the legendary ruler who unified the tribes of northern China in the third millennium B.C. In fact, because of the exponential rise in the number of ancestors as one goes back in time, all Chinese *are* descended from Huang-Ti, if he actually existed. Indeed, he was supposed to have had twenty-four sons, which would have further magnified his genealogical presence. But Huang-Ti was descended from modern humans who had worked their way north from southeastern Asia many thousands of years earlier.

If the movement of people from south to north were the entire story, the genetic history of Asia would be simple. But the occupation of the continent by modern humans was just the first scene of the first act. Since then, continual mixings and movements of people have occurred all across Asia. Often these movements have been related to technological changes. For example, the development of agriculture around the Yellow and Yangtze rivers between 7000 and 6000 B.C. led to rapid expansions of the prehistoric Chinese outward from the northern parts of modern China. Quirks of history, culture, and geography also have left their marks on the genetics of Asian populations. For example, China was conquered several times in its history by tribes of mounted warriors from the north, such as the Mongols and Huns. The newcomers usually were absorbed into the population over time, adding to the genetic diversity of the Chinese. Similarly, as Chinese people moved into other parts of the world, they often mixed with groups they encountered along the way. For example, the genetic evidence indicates that the Tibetans are descended both from northern Chinese who moved south and from central Asian populations. Such population movements have created an incredibly complex mosaic of ethnic groups, languages, and genetic variants throughout Asia.

An important aspect of this history was the development of features that westerners think of as peculiarly Asian — the epicanthic fold, light skin, and relatively flat face. But these features are far from universal in Asia; for example, they are largely absent among New Guineans, Australians, and Polynesians. In contrast, they are especially pronounced in northern Asia, with a secondary distribution in southern Asia that is probably related to the expansion of northerners to the south following the development of agriculture.

For years anthropologists have speculated that these "Mongoloid" characteristics were an adaptation to the cold of northern Asia, particularly at the end of the last Ice Age. This explanation might be right. Short stature, along with the extra layer of body fat often present in Asians, would conserve heat. The flat face would make the nose and lips less susceptible to frostbite. The epicanthic fold would protect the eye from cold winds and perhaps act as a sort of visor for glare from snow.

But the effects of natural selection might be indistinguishable from those of culture. A good analogy from Europe is blond hair and blue eyes. The first babies in Europe with fair coloring, which must have arisen after modern humans moved north from Africa into regions where sunlight is less intense, were probably considered very odd-looking. But some groups must have come to view these features as attractive, or they would not be as widespread as they are today. Mongoloid features are much more common in Asia than is blond hair in Europe, partly because of historical expansions of people with those features. Yet they may very well have originated the same way. Defining Asians as a "race" on the basis of these features therefore makes about as much sense as defining blonds as a "race."

The distribution of physical appearances played out differently in various regions. Japan, for example, was occupied more than 10,000 years ago by a group now called the Jomonese. Judging from their fossils, these people — who appear to have been the first in the world to invent pottery — seem much more closely related to people from New Guinea and Australia than to people from China. Their faces are more contoured than those of the Chinese, as are the faces of all the people who spread from northeastern Africa into southern Asia. A good guess for the origins of the Jomonese, based on the limited genetic data available today, is that they represented an early migration of people north from Australia and southeastern Asia, before the development of Mongoloid features.

Then, about 2,300 years ago, a new group of people moved into Japan, probably from the Korean peninsula. Usually called the Yayoi, these people, who were rice farmers, metal workers, and weavers, had distinctly Mongoloid features. The DNA of modern Japanese shows that extensive mixing occurred between the Jomonese and the Yayoi and that all Japanese today are descended from both groups. Meanwhile the Yayoi pushed the remaining Jomonese to the north, and today in the far north of Japan lives a group known as the Ainu, who almost certainly are descended largely from the Jomonese. They have light skin, features that in some ways look more European than Mongoloid, and more facial hair than their countrymen — they sometimes are called the "hairy Ainu."

The migration of the Yayoi into Japan was part of a much broader movement that carried Asians over more than half the world. From the north they traveled across the Bering land bridge into the Americas. From the southeast they set sail across the Pacific and may have traveled as far as the coast of South America. Asians sailed across the Indian Ocean all the way to Madagascar, where they mixed with Africans. And up and down the length of Asia, they interacted with people who had come out of Africa by a different route.

Wherever people breed more within their own group than outside of it, groups will diverge genetically. This process occurs on all scales, whether between two villages or between two ends of a continent. Because of the great distance from western to eastern Asia, genetic divergence was inevitable. But the extent of this divergence has always been limited by the inevitable human proclivity to interbreed. Eastern Asians were never completely isolated from the people of the Middle East, Europe, and Africa. Mixing between East and West has occurred throughout history, though with different intensities in different times and places.

The earliest interactions undoubtedly occurred along Asia's southern edges. If people could travel east along the shorelines toward Australia, they also could travel west toward Africa. Such movements were probably more difficult once the coasts were settled. Still, plenty of signs point to continued contact between Africa and southern Asia. From Yemen to Papua New Guinea live groups that still look more African than Asian. The Semang in the mountains of the Malay Peninsula, the Andaman Islanders off the coast of Myanmar, and various tribal groups in southern India all have dark skin, curly hair, and wide noses. These features may be adaptations to life in the tropics retained or developed by these populations. But they also may indicate continued movements between Africa and Asia following the initial colonization of Australia.

The second place where East met West was in the north. Soon after modern people moved into the Middle East to stay, their descendants moved northeast to the Asian steppes and then into Siberia. They seem to have been following the great herds of mammoths, bison, horses, and other animals that occupied the vast grasslands of northern Asia. They took with them the advanced tools that characterized the arrival of modern humans in Europe, and they moved quickly.

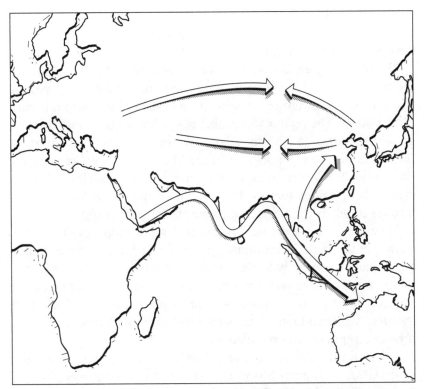

By 65,000 years ago, modern humans had made their way along the shorelines of southern Asia to Australia. They moved north into eastern Asia within the past 40,000 years. In northern Asia, people from the East eventually mixed with groups that had moved from the Middle East into Siberia. More recently, people from East and West have mixed in the oases and steppes of central Asia.

Near the Altai Mountains north of the Tarim Basin, tools characteristic of modern humans have been dated to about 43,000 years ago.

At some later point, these populations of westerners began interacting with people moving north from southeastern Asia. Throughout northern Asia are people with mixtures of European and Asian features. Some groups that occupy central Asia today may descend from these mixed populations. Groups in central Asia, such as the Kazakhs, Kirghiz, and Uighur, clearly have both western and eastern ancestors, as reflected in both their appearances and their DNA.

People and their genes traveled in both directions in northern Eurasia. Geneticists have identified a mutation on the Y chromosome, to take just one example, that is fairly widespread among the Buryat, a group that lives in Mongolia. But the mutation is also found in populations of northern Europe, particularly among the Finns. Given the way the mutation is distributed across Eurasia, people carrying it must have moved west from northern Asia into Europe.

The other major place where people from the East met those from the West was in the Tarim Basin. So far no signs of human occupation that predate the invention of agriculture have been found in this part of the world. The region was probably too barren to attract hunter-gatherers, compared with the steppes to the north and the tropics to the south. But once populations in the Middle East and China began their agriculturally driven expansions, they began to spill into adjacent areas. These groups met in the foothills of the Tarim Basin, where farmers could lay out fields next to the glacier-fed streams.

They met, and they mated. In his analysis of mitochondrial DNA from the Tarim Basin mummies, geneticist Li Jin has found haplo-types characteristic of both East and West, but they are mixed in a particular way. Some eastern haplotypes come from skulls with western features. Some western haplotypes come from skulls with Mongoloid features. Asians and Europeans were producing children with the features and genes of both populations.

Culturally, the division between East and West has been a prominent feature of human history; "never the twain shall meet," wrote Rudyard Kipling in "The Ballad of East and West." Genetically, that division has been a temporary accident of geography and time. "Europeans and Asians have had a brief episode of separation, during which each group acquired enough mutations to be distinguishable from the other," says Jin. "But we're talking about differences that affect a very small proportion of the genes. If we look at the human species as a whole, I don't think it's correct to talk about races. People are so damn similar."

8

Sprung from a Common Source

Genes and Languages

> If we possessed a perfect pedigree of mankind, [then]
> a genealogical arrangement of the races of man would
> afford the best classification of the various languages now
> spoken throughout the world; and if all extinct languages,
> and all intermediate and slowly changing dialects, were
> to be included, such an arrangement would be the only
> possible one.
>
> — Charles Darwin, *On the Origin of Species*

SUPPOSE WE WERE to travel back in time and join the modern humans about to embark on the migration from the Horn of Africa that would take their descendants all the way to Australia. Could we understand what they were saying?

If they built rafts to sail across the Red Sea, they were almost certainly using language to communicate. Felling logs and lashing them together to make a seaworthy vessel is a technological feat far more sophisticated than making stone tools. The migrants would have needed to provision themselves for the journey and rapidly adapt to new environments as they moved. Earlier societies of archaic and modern humans exhibit no clear signs of the technological creativity and social flexibility associated with language. The migration of modern humans from Africa to Australia is different — a journey of such epic scope seems inconceivable without language.

Reconstructing a language spoken more than 65,000 years ago may seem a tall order. Writing was invented a bit more than 5,000 years ago. Everything said before that is gone forever, as irretrievably lost as the footsteps left by the raft builders on the ancient shoreline of the Red Sea.

Yet when these people wanted to refer to the water they were crossing, they may have used a word that sounded something like "aqwa." When they pointed to the flight of a bird overhead, they could have used a word similar to "par." When they gestured toward the ground, they might have said something like "tika." In fact, here's a question in a pidgin version of the language they may have spoken: "Kun mena mana? Kun mena aqwa?" It means: "Who thinks we should stay? Who thinks we should go across the water?"

Most linguists who study the history of language would scoff at this kind of speculation. They say that languages change far too quickly to permit such reconstructions. In the King James Bible, which was published in 1611, the Lord's Prayer begins this way: "Our Father which art in heaven, hallowed be thy name." In eleventh-century English it reads: "Faeder ure thu the eart on heofonum, si thin nama gehalgod." These changes represent just 600 years of language evolution. Over 65,000 years, say most linguists, the odds of any word staying the same are infinitesimally small.

A group of iconoclasts, inspired by the work of the late Joseph Greenberg of Stanford University and his colleague Merritt Ruhlen, disagrees. They say that some words are notably resistant to change — words for parts of the body, personal pronouns such as "me" and "you," and basic features of the environment. Furthermore, subsets of these words are remarkably similar in languages around the world. The word for "knee" in many Australian languages sounds like "bungu"; among the Uighur of Central Asia the word for "kneel" is "bük"; in Swahili to bend down is "bong'oa"; among the South American Guamaca tribe it is "buka"; and the English words "bow" and "elbow" are related to the same root. The word for "woman" or "wife" in the principal language of Kirghistan is "künü," in Cambodian it is "kane," in Zuni, "kanak{ʷ}ayina," and the English word "queen" comes from the same source. These similar-sounding words could not have arisen by coincidence, say these linguists, and they are too separated geographically to have been adopted from language to language all around the world. Rather, the correspondences must mean that they descend from a single original language, which its proponents have termed "Proto-World." And to the extent that any group of people can be identified as the speakers of Proto-World, it would be the first modern humans who left Africa and began colonizing the world.

Darwin was one of many scientists who have wondered about the diversification of languages. He speculated that the development of what he called "the races of man" and the diversification of languages are flip sides of the same coin. At some point, he reasoned, a single language must have been spoken by a single group of humans somewhere in the world. This group then split into two. Over time the physical appearances of members of the two groups diverged, and so did their languages. If this process were repeated many times, the result, Darwin thought, would be a family tree of human races and languages. The original group of humans and the original language would be at the root of the tree, with all of today's groups and languages at the tips of the branches.

As we'll see, this picture of human history has some big problems. But let's assume for the moment that the association between languages and groups is valid. In that case, the distribution of languages in the world should parallel the distribution of genetic variation. If languages diversify the way groups do, then the worldwide patterns of language should match the worldwide patterns of human DNA. Where gaps appear in the genetic record, linguistic analyses could fill them. And both linguistics and genetics could help explain the archaeological record. In other words, by building on one another, the disciplines of archaeology, genetics, and linguistics could be on the verge of a "new synthesis" (in the words of archaeologist Colin Renfrew of Cambridge University) that would shed light on vast spans of human prehistory.

This chapter is the most speculative in this book. A language can change drastically in a single generation, whereas DNA sequences change much more slowly. The history of language is a fantastically complex problem of worldwide scope — this chapter appears in this section only because the languages of Asia and Australia are exceptionally diverse and are central to understanding the overall distribution of the world's languages. In the final analysis, historical linguistics may have little to say about events in the distant past — say, before the invention of agriculture. But the prospect of Renfrew's new synthesis is so enticing that the relationship between genetics and languages cannot be ignored.

*

The origin of language is an area of science largely unencumbered by facts, so hypotheses tend to multiply out of control. This problem is not new. In 1866 the Linguistic Society of Paris banned all discussions of the origin of language from its meetings as a waste of time.

That remains a pretty good policy. We will never know for sure how language arose. Maybe a small group of humans made some sort of conceptual breakthrough that quickly spread to other human populations. Or perhaps various groups were speaking less complicated proto-languages that were swept away by a more sophisticated newcomer. Language might have arisen more gradually through scattered innovations in many different groups. In the absence of any hard evidence, we're left with no more than plausible guesses.

But a couple of observations can be taken as starting points. The first group of modern humans to sail across the Red Sea must have been speaking a single language. They needed language to communicate, and the group was probably small enough that they all spoke the same language. Similarly, the first group of modern humans to sail across the Timor Sea from modern-day Indonesia to Australia must have been speaking a single language, for the same reasons.

Those languages were probably not the same. For most of human history, languages have tended to be localized, spoken by particular groups in particular places. If a group split, the languages spoken by the secondary groups tended to diverge, just as Darwin supposed. In this respect, the divergence of languages does bear some similarity to the divergence of DNA. As long as a group of people is intact, they share both their genetic mutations (through interbreeding) and their language (through conversation). But once a group divides, innovations in language have few ways of getting from one group to the other. In each group new pronunciations come into vogue, old words are replaced by new ones, and new ways of combining and inflecting words take hold. As a rule of thumb, say historical linguists, if two groups speaking the same language are separated for more than about a thousand years, their languages will change so much that they will be mutually incomprehensible.

Modern humans probably took at least this long to travel from Africa to Australia. Even if the people who reached Australia were descendants of those who left Africa, their language probably would have changed enough to not be understood by the people they left behind. And that doesn't even consider the inevitable complications of the migration into Asia, in which multiple groups coming out of Africa could easily have mixed.

But even if the languages spoken at the beginning and end of the migration to Australia differed, linguists could probably tell they were related. Individual words would have changed but would still bear some resemblance to the originals. The word for "heaven" in eleventh-century English, for example, was "heofonum" — different but not unrelated. Similarly, grammatical aspects of the two languages — for example, whether the object of a sentence follows or precedes the verb — would reveal commonalities. These are known as genetic correspondences (a happy choice of words) and they indicate that the two languages are descended from a common ancestor.

The classic example of genetic correspondences comes from western Asia and Europe. For more than two centuries, scholars have known that most of the languages spoken in a broad swath from Britain to India are derived from a single original language. These languages belong to what is called the Indo-European family, and they are spoken by more people than are the languages of any other family in the world. They include Indic languages such as Hindi and Urdu, Iranian languages such as Farsi and Kurdish, Slavic languages such as Russian and Serbo-Croatian, Romance languages such as Italian and French, Germanic languages such as Norwegian and English, and Celtic languages such as Irish. Today many of these languages seem to bear only the slightest resemblance to each other. Yet most linguists believe they all are descended from a single ancestor language spoken by a small group of people — probably a few thousand — living in a relatively small area, perhaps just a few thousand square miles.

A remarkable amount is known about the original speakers of what is called Proto-Indo-European. They raised cattle, kept dogs as pets, used bows and arrows in battle, and worshipped a male god associated with the sky. All this information comes from linguistic reconstructions of their language. If a number of Indo-European languages have words that derive from the same root word, linguists can conclude that Proto-Indo-European contained that word. What's more, by studying how individual words have changed in different Indo-European languages, linguists can work out an earlier form from which all the derived forms descend. For example, the word "father" in English, "padre" in Spanish, "pater" in Latin, and "pita" in Sanskrit all come from the Proto-Indo-European form *pəter* (the asterisk means that the word is a reconstruction). The Proto-Indo-European word for sky was *dyeus — Ju* in Latin, leading to the Latin word "Ju-pater," or sky father. Dictionaries of Proto-Indo-European now contain several thousand words.

Yet for all that is known about the group that spoke Proto-Indo-European, linguists cannot agree on the two most basic facts about its existence: when and where these people lived. Proto-Indo-European contains many words for different kinds of trees, so they probably lived in or near wooded areas. They also had many names for birds and made clothes from wool. Unfortunately, these clues could point to almost anywhere in Asia or Europe. At the moment there are two leading hypotheses about the Proto-Indo-European homeland. One is that the language was spoken by a nomadic people who lived just north of the Black and Caspian seas about 6,000 years ago. These people may well have been the first to domesticate the horse, which would have given them a great advantage in warfare and perhaps in trade. As their descendants — or perhaps just their way of life — spread across Europe, so did their language. Over succeeding millennia, this language evolved into the Indo-European languages known today.

The second hypothesis advocates an earlier spread of these languages. According to this idea, the first farmers to spread out of the Middle East into western Asia and Europe, beginning about 9,000 years ago, spoke Proto-Indo-European. As the farmers mixed with the indigenous peoples of these areas, and as indigenous people learned how to farm, Indo-European languages replaced the earlier tongues.

Both hypotheses have their problems. Clear evidence for the domestication of the horse 6,000 years ago is hard to find. People were certainly eating horses, but they may not have been riding them until much later. Nor is there any evidence for widespread conquests within Europe and western Asia by mounted warriors, which is one popular idea about how Proto-Indo-European spread. At most, some Europeans and west Asians might have adopted some elements of steppe culture during the fourth and third millennia B.C., but whether these changes involved language is unknown.

The possible spread of Indo-European languages by advancing waves of farmers also has problems. No evidence indicates that the early farmers spoke these languages. Indeed, judging by ancient place names in Turkey, they may well have spoken a non-Indo-European language. Similarly, Proto-Indo-European seems to have had words for copper, cart, and the domesticated horse, yet these were unknown among the first farmers of Turkey.

Today the people of the world speak more than 5,000 languages, all of which arose through three basic mechanisms. The first is colonization, in which modern humans entering a new area introduce their language. The second is divergence, in which a language spoken by different groups changes over time. The third is language replacement, in which a group of people adopts a new language introduced or imposed by another group.

These three processes have created the tantalizing patterns of linguistic structure that exist in the world today. Linguists know that many languages are related through descent from a progenitor such as Proto-Indo-European. But languages and the families to which they belong seem related on multiple levels, and many of the relationships are obscure. Relationships that seem obvious to one linguist are nonexistent to another. And because linguists cannot agree on the criteria to be used to define those relationships, they tend to engage in an activity for which they are well known: they argue.

In general, two schools of thought exist. Some linguists are lumpers; they group as many languages as they possibly can into single families. Others are splitters; while they acknowledge some similarities between languages, they contend that these arose in other ways, not from common descent.

Over the next several pages I'm briefly going to describe nineteen major language families spread across the world. In my descriptions I generally lump languages together rather than splitting them apart. The possible connections between languages and genetics are clearer when dealing with fewer language families. Also, the process of working out relationships among languages and language families is ongoing, since many languages have not received the kind of study needed to determine their links with other languages. Of course, any classification will offend some linguists. But good reasons exist for grouping languages into the nineteen families shown on the accompanying map and described below.

I also concentrate on broad patterns rather than on the details. People and their languages are always on the move. Even before the colonizations of the past few centuries, many languages were spoken far from their homelands, whether because of trade, war, or migration. A

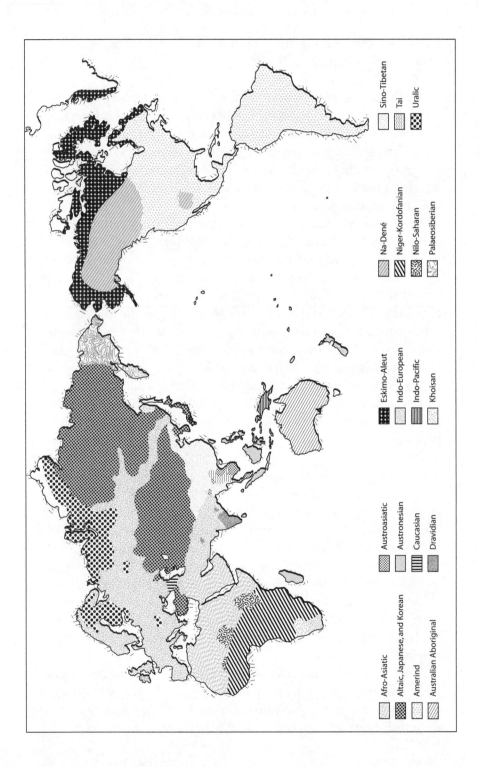

Afro-Asiatic		Austroasiatic		Sino-Tibetan
Altaic, Japanese, and Korean		Austronesian		Tai
Amerind		Caucasian		Uralic
Australian Aboriginal		Dravidian		

Eskimo-Aleut		Na-Dené
Indo-European		Niger-Kordofanian
Indo-Pacific		Nilo-Saharan
Khoisan		Palaeosiberian

completely accurate linguistic map — showing exactly what is spoken where — would be hopelessly complex. And a few of the world's languages, such as Basque, don't fit comfortably into any language family, hinting at even deeper layers of complexity.

AFRICA

The languages spoken south of the Sahara are diverse, as are the people who live in this part of the world. Yet in 1963, Joseph Greenberg proposed that these languages could be grouped into three broad families. Although his classification originally met with vehement criticism, today almost all linguists agree with it.

The Bushmen of South Africa and their pastoralist cousins the Khoi-Khoi speak what are termed *Khoisan* languages. These are the only languages in the world that rely heavily on a wide variety of complicated click sounds. At least five kinds of clicks are made with various parts of the tongue, throat, and mouth. These clicks can be spoken on a rising or a falling tone, on an in-breath or an out-breath. They make the Khoisan languages incredibly difficult to learn after childhood. By the same token, listening to a fluent Khoisan speaker is a wonderful experience, since the sounds are so different from those of all other languages.

Some linguists have speculated that these clicks identify the Khoisan languages as the world's oldest, with the clicks then being lost in the languages that spread out into the rest of the world. This theory corresponds well with the very old mitochondrial and Y haplotypes found in many Bushmen. But another possibility is that the clicks entered the Khoisan languages after they diverged from earlier languages and never spread more widely.

The Khoisan languages have one major anomaly. In Tanzania, more than a thousand miles from the Bushmen and Khoi-Khoi, live two small groups of people, the Hadza and Sandawe, who speak languages that also have many click sounds. Anthropologists do not yet know whether these groups are descended from southern Africans who

Opposite: Almost all of the world's 5,000 languages can be sorted into nineteen families. The distributions of these families and of genetic variation bear remarkable similarities, though many of the relationships between genes and languages remain obscure.

migrated to eastern Africa or whether they represent the remnants of a Khoisan-speaking people who once lived throughout eastern and southern Africa.

Most of the people of western, central, and southern Africa speak languages belonging to the *Niger-Kordofanian* family. By far the largest number of languages in this family belong to the Bantu subfamily, which spread from present-day Nigeria starting about 2,500 years ago.

Languages belonging to the third major family in sub-Saharan Africa, the *Nilo-Saharan* family, are spoken in areas scattered to the north and east of the Niger-Kordofanian languages. Some linguists have suggested that the Nilo-Saharan and Niger-Kordofanian families have enough similarities to be combined into a single family, a grouping that would encompass most of the languages spoken by sub-Saharan Africans.

THE MIDDLE EAST

Most of the people of the Middle East and northern Africa speak languages belonging to the *Afro-Asiatic* family. This family consists of six subfamilies, the largest of which is the Semitic branch, which includes Hebrew, Aramaic, and Arabic — the languages of the Old Testament, the New Testament, and Islam, respectively.

As with the Indo-European languages, the Afro-Asiatic languages seem to have descended from a single language originally spoken in a small region. But Proto-Afro-Asiatic appears to be older than Proto-Indo-European. The latter was clearly spoken after the invention of agriculture, as attested by the many words for domesticated plants and animals in the reconstructed language. Proto-Afro-Asiatic appears to have predated the invention of agriculture. No one knows for sure, but perhaps it was spoken by the first people to establish permanent settlements, the Natufians, and as their way of life spread, so did their language.

ASIA AND AUSTRALIA

The continual movements and mixings of peoples throughout Asia have produced an extremely complex pattern of languages. Some of the loudest debates about language families center on Asia.

Probably the oldest languages in the region are those of Australia and New Guinea. When Europeans first arrived in Australia in 1788, about 260 *Australian Aboriginal* languages were being spoken by the inhabitants of the continent. (More than 160 of these languages are now extinct, and only about 20 are spoken by sizable numbers of people.) Even more languages, roughly categorized as *Indo-Pacific*, were being spoken in New Guinea. Even today the several million people who live in the interior of the island, which is just a little larger than Texas, speak almost 700 languages. Most of these probably are descended from languages spoken by the original settlers of New Guinea. Yet many are so different that some linguists question whether they are even related to the others.

In southern India, most people speak languages belonging to the *Dravidian* family. The language spoken by the ancient Elamites, who built a powerful state in southern Iran in the third millennium B.C., may also have belonged to this family. One possibility is that Dravidian languages were once widespread between the Middle East and southern India, perhaps as the result of an eastward expansion of farmers from the Middle East. According to this hypothesis, these Dravidian languages later were replaced in Iran and northern India by the expansion of Indo-European languages.

The language families of mainland and insular southeastern Asia are the most tangled in the world. *Austroasiatic* languages, including Vietnamese and Cambodian, are spread throughout southeastern Asia. *Tai* languages are widely spoken in Thailand and Laos, with extensions into Burma and China. *Austronesian* languages are spoken across more than half the world, from Madagascar to Taiwan to Easter Island to Hawaii, because of an expansion of farmers from southeastern Asia over the past few thousand years.

Some linguists believe that all three of these language families are part of a larger group called *Austric*. If so, Proto-Austric may have been one of the earliest languages spoken in Asia, as modern humans began working their way north from Australia and New Zealand. But in general the languages of southeastern Asia have not yet received enough scholarly study to support or reject this hypothesis.

Farther north, the majority of people in China speak languages belonging to the *Sino-Tibetan* family. There are eight major Chinese dialects, though some linguists say these should be termed languages, since they are not mutually intelligible. For example, the residents of Shanghai speak a dialect that cannot be understood in Beijing, though most people in Shanghai also speak Mandarin, the official Chinese language. However, the written words of the various dialects are the same, because they are based on ideographs rather than an alphabet. People speaking different dialects therefore can read the same newspaper, but if they were to read the newspaper out loud, the words would sound different.

In northern Asia the most common languages are those of the *Altaic* family, which includes the Turkic and Mongolian languages. The Turkic languages are widespread, spoken all the way from Turkey to northeastern Asia. They also are remarkably similar, possibly because the nomadic peoples who spread these languages have maintained a dense network of linguistic connections across Asia. Another language family, known as the *Uralic,* spreads across the north of Asia and back into Europe. Finnish, for example, is a Uralic, not an Indo-European, language.

The proper classification of Japanese, Korean, and the languages spoken by the Ainu of northern Japan remains controversial. Some linguists group these languages with the Altaic family, though with a heavy influence from Chinese. Others say they belong in categories by themselves. Since I've chosen to be a lumper rather than a splitter, I've put them with the Altaic languages.

The *Palaeosiberian* language family is a valuable reminder of the inevitable complexities of linguistic classification. The grouping consists of several languages with no obvious relation to each other spoken by a few thousand people scattered throughout northeastern Siberia. The description of these languages as a "family" is a product only of tradition and convenience. In their uniqueness, the Palaeosiberian languages resemble several other small isolates scattered across the world, such as Burushaski, a language with no known relatives spoken by more than 20,000 people in Kashmir, India, and Pakistan.

EUROPE

Most of the languages of Europe are *Indo-European,* with a couple of intriguing exceptions. One of these exceptions is Basque, which is spoken by about a million people in the mountainous regions near the border of Spain and France and appears to be a remnant of the languages spoken in Stone Age Europe. The genetic background of the Basques is also different from that of other Europeans. What appears most likely is that the people who spread Indo-European languages bypassed the Basques' mountainous homeland, because it was either too difficult to farm or too difficult to conquer. Though the Basques have certainly mixed with other Europeans, they may be the people most directly descended from the original modern humans of Europe.

A few other non-Indo-European languages survived into the age of writing in Europe and then went extinct. For example, Etruscan, a language spoken widely in Italy before the rise of Rome, left records that still have not been completely deciphered. Hungarian, in contrast, is a Uralic language introduced to Europe when the nomadic Magyars invaded Hungary from Russia in the ninth century A.D. Several other non-Indo-European languages spoken in eastern and central Europe date to similar invasions and conquests.

Another rich area for languages in Europe — really at the intersection of Europe, the Middle East, and Asia — is the Caucasus Mountains. In a tiny region between the Black and Caspian seas, people speak a breathtakingly diverse collection of languages. In addition to the Indo-European, Semitic, and Turkic languages that have been introduced to the region in the last few thousand years, about forty languages are spoken that belong to an ancient family known as *Caucasian*.

THE AMERICAS

For several decades Joseph Greenberg contended that all of the languages spoken by Native Americans can be divided into three broad families, a grouping that has not yet won wide acceptance. The most recent family, according to Greenberg, encompasses the *Eskimo-Aleut* languages spoken by the people of the far north. The second oldest family consists of the *Na-Dené* languages, spoken mostly in northwestern North America, although Navajo and Apache, in the southwestern United States, are also Na-Dené. The largest, oldest, and by far most controversial language grouping in the Americas is what Greenberg termed the *Amerind* family. It includes all the other languages spoken by the native peoples of North and South America, from the Algonquian languages of the north to the Andean languages of Tierra del Fuego.

Darwin thought that if the connections among all the world's languages could be traced, then so could human history. As he wrote in *The Descent of Man*, "If two languages were found to resemble each other in a multitude of words and points of construction, they would be universally recognised as having sprung from a common source."

Based on my descriptions of the world's language families, Darwin's conjecture looks fairly good. The distribution of languages does bear some striking resemblances to what is known both from archaeology and from studies of genetic variation. These correspondences are best for the most recent events. For example, the expansion of agriculture is closely related to both languages and genes. The Indo-European, Afro-Asiatic, Sino-Tibetan, Bantu, and Austronesian languages all appear to have spread from centers of agricultural development to adjoining regions. The Altaic languages might have spread not so much with farmers as with pastoralists who drove their herds across the Eurasian steppes.

But if modern humans have common origins, then at least some of these language families should themselves be related. These higher-level connections are extremely controversial, given that many linguists can't even agree on which language families are valid. But some intriguing suggestions have been made. For example, various groups of linguists have proposed a connection among language families stretching from northwestern Africa across northern Europe and Asia to northeastern Siberia. This "Eurasiatic" or "Nostratic" macrofamily would include the Indo-European, Uralic, Altaic, and Eskimo-Aleut families (different formulations group them somewhat differently). For example, in many of the languages in these families, the first-person pronoun starts with an *m* sound (as in "me" in English) and the second-person pronoun starts with a *t* sound (as in "tu" in French or Latin). Many other similarities exist among these languages — so many that linguists have begun to put together dictionaries of Proto-Eurasiatic root words. But Proto-Eurasiatic has not been studied enough to even begin to guess where or when such a language might have been spoken.

A connection among these languages makes sense genetically, for all have at least a plausible link with the Middle East. This connection may date back to the emergence of modern humans from Africa, or it could be more recent, maybe involving the social changes that paved the way for agriculture. Perhaps further reconstructions of Proto-Eurasiatic will reveal something about the people who spoke such a language and thus provide some clues about where and when they lived.

Beyond Proto-Eurasiatic lies the ultimate objective of Proto-World advocates: the demonstration that all the world's languages are related to each other. But except for provocative lists of similar-sounding words, progress toward that goal has been slow. Some linguists have proposed a possible link among Basque, the Caucasian languages, the Sino-Tibetan languages, and the Na-Dené family of North America. But most scholars, including those sympathetic to the notion of Proto-World, believe it is premature to begin reconstructing a language so old that it could have given rise to both Basque in western Europe and Na-Dené in the Americas. The similar-sounding words among languages could be just coincidence, they say, not a reflection of a common origin. Besides, there are other ways in which languages come to resemble each other. Indeed, these mechanisms could turn out to be the ultimate showstopper in efforts to find connections among ancient languages.

Say that people learned and spoke only one language, that of their parents, their whole lives. In that case languages would be just like genes. People would acquire a language from their parents and pass it on to their children in an unbroken chain.

But languages don't work like that. Some children learn a different language from that of their parents and eventually give up the first. Because every language changes over time, the language people learn as infants can differ from the language they speak in old age. Sometimes a new language can sweep over a part of the world and quickly replace the languages spoken there, as happened in parts of Africa, Asia, and the Americas following European colonization.

Even if a language is not replaced, it may adopt sounds, words, and even grammatical structures from another language. My favorite example is the word "okay." Anywhere in the world, if people know a single word of English, it's "okay." "Okay," says the taxi driver in Shanghai when you tell him where you want to go (even if he has no idea what you said). "Okay," says the Bushman in Botswana when you've mimicked a complex dance step. Eventually the dictionaries of many languages will include a word sounding something like "okay" because of its ubiquity in everyday life.

A linguist working centuries in the future could easily conclude that "okay" must have been one of the first words uttered by modern humans, since it is so widely distributed. In fact, "okay" is the product of a bizarre set of accidents that occurred in the 1830s and 1840s in the United States. A fad arose of abbreviating various stock phrases in semi-humorous ways. "OK" was the abbreviation for the misspelled phrase "oll korrect" (another such abbreviation was NG for "no go"). But "okay" might never have survived if not for another quirk of history. In 1840 Martin van Buren ran for president. His nickname was "Old Kinderhook," and his supporters started a booster organization called the OK Club. Soon the word became so popular that it was used simply to mean "that's good," and the usage spread.

Actually, linguists have many methods for detecting borrowing among languages. But these methods become less effective the farther back one looks, and borrowing has undoubtedly occurred as long as people have been talking to each other. Borrowing must have been especially common when the speakers of two languages had extensive contacts, say as part of a trade network or through intermarriage. In such circumstances, people would have had to become bilingual, so one language could more easily influence the other.

The linguistic maps of Australia and New Guinea are a good example of how this process might have worked. Many of the languages spoken by the aborigines of Australia are quite similar, suggesting descent from a common source. That source may well have been the original language spoken by modern humans in Australia, since the continent's languages bear none of the telltale signs of having been replaced in the past. But if Australia's languages were all descended from a single language spoken 65,000 years ago, they should have diverged so drastically that they would have very few connections. Instead, they appear to have undergone both divergence and convergence, with exchanges of words and grammar ensuring that they did not become greatly separated.

Such exchanges make sense in the case of Australia. Aboriginal groups have always been in contact with one another across the continent's dry, open plains. These contacts could have kept the languages from diverging in major ways.

In New Guinea, groups have been more isolated in the mountainous interior of the island, and their languages have diverged much more dramatically. Groups in New Guinea still interact and communicate, often using second languages that are more widespread. But the geographic and ecological separation of groups seems to have encouraged language divergence rather than convergence.

If convergence has been a common feature in the history of languages, then all bets are off. Languages could be similar to each other not because they are descended from a common ancestor but simply because they are spoken in adjoining areas. Furthermore, adjacent languages could establish chains of convergence, where each gradually becomes more like its neighbors. Over time, this convergence process could stretch across entire continents. Perhaps this explains the similarities seen in languages across northern Europe, or even some of the similarities seen within language families.

The possibility of convergence does not ring a death knell for the study of language families. The descent of the Indo-European languages from Proto-Indo-European, for example, is so well established that the overall process is beyond dispute. But the intermediate steps become more complicated. The path from a proto-language to today's languages becomes much more obscure if convergence is common. In many cases, the effects of common descent and of borrowing may be impossible to separate.

Convergence also may set an absolute limit on how far into the past we can track connections between languages. Beyond some period of time, the traces of a common origin may be lost among the tangled skeins of words. If so, the hopes of those seeking a single original language will be dashed. That language, if it ever existed, may be gone forever.

Where does this leave the "new synthesis" — the idea that archaeology, linguistics, and genetics are on the verge of producing a coherent and fairly complete picture of human prehistory?

Such a synthesis may still emerge, but it is not here yet. Only as more data are gathered in each field will the picture become clearer. Additional genetic information is needed to trace the movements and mixings of people with greater accuracy. Genetic data will suggest promising archaeological investigations. Linguists need to investigate the links between languages in greater detail — an especially urgent task, since thousands of languages are now on the verge of extinction.

Science works by proposing explanations and then seeing if they can be demonstrated to be wrong. If the explanations work well in many cases and are not obviously contradicted, they gradually will be accepted. If the explanations forecast patterns not previously noticed, then they will be held with even greater confidence.

Archaeology, linguistics, and genetics all suggest explanations about our history. Any given one could be wrong. Yet the effort is sustained by the most straightforward of observations: just one human history has occurred. With luck, good planning, and persistence, maybe the explanations coming from these three fields eventually will converge on the same story.

IV

Europe

9

Who Are the Europeans?

Purity of race does not exist.
Europe is a continent of energetic mongrels.

— H. A. L. Fisher, *A History of Europe*

HEADED WEST FROM LONDON across the Salisbury Plain, a jet-lagged American might be forgiven for imagining, just for a moment, that he is driving through the American Midwest rather than England. In every direction, freshly plowed fields roll in broad waves to the horizon. Cattle wander through bright green pastures. Occasionally the rich tang of manure — redolent of life, commerce, and decay — wafts from nearby pens.

Suddenly a jumble of stones rises not 200 feet from the busy roadway. It's Stonehenge, the most famous prehistoric monument in all of Europe. The site straddles a gentle hill overlooking close-cropped fields and distant lines of trees. It's a bizarre anachronism in the midst of so much farmland. In 1135 Geoffrey of Monmouth wrote that Merlin had magically transported these stones here to create a monument to four hundred British chieftains killed by the German warrior Hengist. The idea does not seem so absurd when you see those wild, primitive stones plunked down in England's cultured landscape.

"Many people tell me that they thought it was going to be much bigger," says the caretaker stationed in front of the monument to make sure visitors don't stray from the paved path. Yet the stones seem breathtakingly large to me. The tallest is four times my height, perfectly vertical and so smooth it would be impossible to scale. The horizontal stones perched atop the massive uprights are as big as small trucks and undoubtedly heavier. The thought of people lifting those stones with their bare hands makes your bones ache.

Every June 21, a peculiar ritual occurs at Stonehenge. Beginning the previous evening, people from all over the world begin gathering at the monument. Many are dressed in white robes and sandals. They belong to groups with names like the Ancient Order of Druids or the Church of the Universal Bond. As the sun comes up on the solstice, the longest day of the year, an unearthly sound rises from the congregation. People chant and sing. Some dance or kiss. Others watch the sun in silence (if it's not cloudy), seeking to capture the mystical emanations of long-dead shamans.

Well, people can think whatever they want to about Stonehenge. But if the modern-day pagans of the Salisbury Plain think that Stonehenge has anything to do with the druids — the priests of ancient Britain when the country was invaded by the Romans in the first century B.C. — they are misinformed. The monument was built thousands of years before the druids existed. By the time of the Roman invasions, Stonehenge was already in ruins.

We actually know little about how prehistoric people used Stonehenge and the hundreds of other monuments scattered throughout England and northwestern Europe. Many of the monuments stand guard over deposits of human bones, so they must have played some role in funerary rites. And many of the monuments — whether great arcs of stone enclosing entire villages, artificial hills large enough to dominate the landscape, or simple burial cairns — take the form of circles, which must have had a special significance to the builders. But did the circles represent the inclusion of all people or the exclusion of some? Did they connote the wholeness of human life or the separateness of the sacred? No one knows.

Archaeologists who have studied the great prehistoric monuments of northwestern Europe are fairly confident about one thing. The monuments seem to have a deep and abiding connection with the most pedestrian aspect of their settings — the fields and pastures that surround them on every side. Many of the monuments were built by people who were making the transition from hunting and gathering to farming. As such, they reflect not only a changing relationship between people and their surroundings but also a profound mental shift, a new way of thinking about time and space. And in that respect they provide a vital part of the answer to the question posed by this chapter: Who are the Europeans?

By the time Stonehenge began to be built, about 5,000 years ago, modern humans had been living in Europe for almost 35,000 years. The earliest modern humans in Europe almost certainly came from the Middle East, traveling either counterclockwise around the Black Sea basin through the Caucasus Mountains and the Ukrainian plains or clockwise through Turkey, Greece, and the Balkan Mountains. Almost all of the mitochondrial and Y-chromosome haplotypes found among Europeans today derive from haplotypes that still exist in the Middle East. The two other possible routes into Europe — by boat across the Strait of Gibraltar or by land from somewhere in Asia — seem much less likely, given the genetic evidence.

Those first modern humans were not moving into uninhabited territory; the Neandertal people had occupied the continent for hundreds of thousands of years. The Neandertals had a culture well suited to the harsh climate of that time. They hunted big game using stone tools flaked from carefully prepared cores. They cared for those who were sick and crippled. At least on occasion, they buried their dead.

But their culture was no match for that of the newcomers from the Middle East. These modern humans either brought with them or developed during their initial movements into the Middle East and Europe a way of life more sophisticated than that of any other humans alive at the time. The stone tools they made were specialized for particular tasks, such as scraping hides. They fashioned bone, ivory, and antler into projectile points, awls, punches, and needles. They developed regional toolmaking traditions, as if the modern humans of Europe were dividing into groups, each with its own distinctive style.

They also created art, something the Neandertals had never done. This artwork is so accomplished that it still has a profound emotional power. The cave paintings of southwestern France, the animal sculptures of central Europe, the carved figurines of women found throughout the continent — all exhibit the visceral connectedness of great art, the ability to reveal to its viewers something about how the artist perceived the world. In the hunter-gatherers of Stone Age Europe we can recognize something of our own minds, which is not easy to do with the Neandertals they replaced.

Why was the culture of these modern humans so much more sophisticated than cultures elsewhere in the world? At that time modern humans had already expanded to Australia and southeastern Asia, southern Africa, and India, replacing the archaic humans they met on the way. Yet the modern humans in those parts of the world were not making the same elaborate tools and artwork as were the Europeans.

For the past several centuries, the answer to this question has seemed obvious to many Europeans: the people of Europe were simply more biologically advanced than others. As the European explorers of the sixteenth and seventeenth centuries encountered people with different appearances and less advanced technologies, they began to rank human groups, always putting themselves at the top. The differences among cultures appeared to Europeans so stark and absolute that biology had to be the cause.

Modern genetics has thoroughly discredited these notions. The genes brought to Europe by modern humans were no different from those of the people who remained in the Middle East and Africa. Over time the frequency of particular genetic variants changed. For example, the genes responsible for skin color must have altered as Europeans adapted to more northern climates. Europeans also developed a suite of somewhat distinctive facial characteristics, as often happens when groups of modern humans become genetically separate from their neighbors. But the genetic affinity between Europeans and people elsewhere is unmistakable.

Consider the people of India. Physical anthropologists traditionally have classified Indians as "Caucasians," a term invented in the eighteenth century to describe people with a particular set of facial features. But this classification has never sat particularly well with some Europeans, who were offended by being lumped with the dark-skinned people of the subcontinent. Gradually a kind of folk explanation emerged, which held that several thousand years ago India was overrun by invaders from Europe. These light-skinned warriors interbred with the existing dark-skinned populations to such an extent that the Indians acquired European features.

Recent studies of mitochondrial DNA and the Y chromosome have revealed a different picture. Incursions of people from Europe into India have certainly occurred, but they have been less extensive than supposed, and genes have flowed in the opposite direction as well. The physical resemblance of Europeans and Indians appears instead to have resulted largely from their common descent from the modern humans who left Africa for Eurasia. In fact, this observation applies in other parts of the world as well. Anthropologists have often speculated about the seemingly "European" features of a number of groups in Asia and even the Americas, groups such as the Australian aborigines or the Ainu. Now the connection among these groups is becoming clearer: they all mixed relatively little with groups that acquired the typically "Mongoloid" features of eastern Asia. Thus the "Caucasoid" features of various peoples around the world may simply reflect the features of the northeastern Africans who gave rise to all the people of Europe and Asia.

To account for the cultural achievements of Europeans, both in the Stone Age and more recently, we must look beyond the discredited genetic explanations of the past. We have to consider instead the factors that made Europe unique, beginning with its unparalleled geographic and biological diversity. Although it is really just a peninsula of Eurasia, Europe is blessed with one of the most varied and abundant physical settings on earth. To the south it borders the placid waters of the Mediterranean, providing ready access to the Middle East and Africa. To the east it merges with the vast steppes of Asia, whose people have had a major influence on the continent. To the west it is bounded by the Atlantic, which serves not only as a route to the rest of the world but also as a critical buffer on climate. Because of the oceanic currents that flow across the Atlantic, Europe is remarkably warm despite its far northern location. This warmth influences even the most northern fringes of the continent, where the forests give way to subarctic plains.

Internally, Europe is even more diverse. It has high mountains, fertile prairies, extensive forests, and vast swamps, and it is threaded by rivers that provide an easy way to move from one ecological niche to another. Because of its many bays and gulfs, Europe has much more shoreline per unit of area than does any other continent. Also, unlike Asia and Africa, it does not have large areas of homogeneous terrain, so people in nearby regions could develop quite different ways of life. Interactions among these groups, whether trade, cultural diffusion, or warfare, could then drive social change.

Another factor that contributed to the cultural efflorescence of Stone Age Europe was the climate. Between 40,000 and 30,000 years ago, when modern humans were replacing Neandertals, the continent was significantly colder than it is today. The summers were warm, but the winters were brutal, with many weeks of subfreezing temperatures and heavy snowfall. This climate must have posed severe challenges to modern humans, whose long limbs are more adapted to the warmth of the tropics. Yet it also provided tremendous opportunities. During this period much of northern Europe consisted of vast subarctic grasslands. Huge herds of game wandered the plains, including reindeer, wild horses, bison, mammoths, and woolly rhinoceroses. Modern humans clearly organized their lives around these animals. They made their camps along the rivers, where herds of game descended from the high regions toward the coastal plains as winter approached. Many of their tools were designed to process the products of the hunt. The openings of the caves where they lived often faced south, so that on cold days the people could be in the sun while working at their tasks.

Still, the weather could be cruel as well as generous. A few thousand years after the death of the final Neandertal, which probably occurred about 29,000 years ago, Europe entered a climatic crisis. The weather became even colder; glaciers pushed south until they were within a hundred miles of modern-day Stonehenge, Amsterdam, and Moscow. Average temperatures were as much as 20 degrees Fahrenheit lower than they are today. So much ocean water was locked up in the polar ice caps that the English Channel dried up and England became part of the continent.

During the height of the Ice Age, between about 20,000 and 16,000 years ago, modern humans gave up on northern Europe, abandoning what is now Britain, northern France, the Low Countries, Germany, and most of Poland. Small groups may have wandered into those areas during the summers, but they left no traces of their visits. Europeans retreated into the warmer areas around the Pyrenees and the Balkans and north of the Black Sea.

This period of intense social crowding was one of great innovation. Artwork flourished among the dense populations of southern Europe. New technologies were developed, such as spear-throwers that allowed hunters to launch projectiles toward their prey with great force. Groups seem to have heightened their cultural distinctions from each other, as if they were marking off separate territories for themselves.

Then the glaciers slowly began to retreat. By 13,000 years ago, people had moved back into northern Europe, including England and northern Germany. At first they undoubtedly resumed their hunting of big game. But as the warming continued, the forests expanded, swallowing up the grasslands of the north. Forests have much less game than do the wide-open spaces of the subarctic plains, so they can support fewer hunter-gatherers. The huge herds of animals began to disappear, in part perhaps because of overhunting by humans. The polar ice sheets melted and sea levels rose, encroaching on the plains from the north. By 8,500 years ago Britain was again an island, cut off from the mainland by the rising waters of the channel.

As the forests spread, people were forced to depend more on fishing, hunting of birds and small animals, and gathering. Archaeological investigations have shown that populations in northern Europe declined as the plains contracted, and people lived in smaller groups that were farther apart. The creation of artwork fell off drastically, replaced by geometric engravings and paintings on pebbles and bones. Even the stone tools became simpler, less varied, and less carefully made. By 10,000 years ago Europe had entered a period of cultural torpor. The conditions that had led to the flourishing of Stone Age culture in Europe were gone.

Popular images of cavemen in cartoons, movies, and books actually describe a very small fraction of the modern humans who lived on the earth before the invention of agriculture. In these depictions, cavemen are typically light-skinned, heavily bearded hunters of big game. They wear animal skins and huddle around fires for warmth. In other words, they are the people of Stone Age Europe, though without the complex culture and languages we knew they possessed.

If pressed, most Europeans would probably agree that they must be descended, in some vague way, from these prehistoric hunter-gatherers. But some scholars have another hypothesis about the origins of Europeans. They cite genetic evidence to argue that Europeans are descended at least in part from farmers who migrated out of the Middle East after the invention of agriculture — that is, within the past 10,000 years. This idea, which was proposed about thirty years ago, has been one of the most controversial applications of genetics to the study of history. And at the center of the debate is the man who has done more than anyone else to define and promote the field, a professor of genetics at Stanford University named Luca Cavalli-Sforza.

Though Cavalli-Sforza has lived in the United States for more than thirty years, his ties to the Old World remain strong. Born in Italy in

1922, he enrolled in medical school at the University of Pavia at the age of sixteen, largely because he was fascinated with microscopes. "It turned out to be a very lucky choice," he told me when I visited him in his offices at Stanford. "Had I not gone to medical school, I would have been conscripted at the beginning of the war." While childhood friends were serving in the Italian army, Cavalli-Sforza was skipping lectures to spend more time on research projects in the lab. Initially he focused on bacteria. He still is well known for his contributions to the discovery that microbes have sex — or at least can engage in the kinds of genetic exchanges involved in sexual reproduction.

He practiced medicine for a year after graduating in 1944 but disliked it intensely. So he found a part-time faculty position at the University of Parma and resumed his research on bacterial genetics. Then in 1951 a chance remark by a student redirected his research toward human genetics. "Throughout my life I've been very lucky at finding good students," he said. "This particular student was also a priest. He mentioned to me that there were some data he thought would be of interest for human genetics." For more than three centuries the Catholic Church had collected information on births, deaths, and marriages in many Italian parishes. These data were ideal, Cavalli-Sforza realized, for studying a particularly contentious issue in twentieth-century genetics — the role of genetic drift in evolution.

Genetic drift is the neglected stepsister of natural selection. When a genetic variant makes an organism more likely to produce offspring that survive into the next generation, the variant can become more common in the gene pool. That's natural selection. But a genetic variant also can become more common purely through luck. If an organism with a particular genetic variant just happens to have many offspring, and if they inherit that variant, it can rise in frequency whether it is selected for or not. Genetic drift is a molecular roll of the dice, the product of genetic chance.

In the 1950s many geneticists disparaged this random element in evolution. With little understanding of the actual mechanisms involved, they believed that natural selection would almost always squelch genetic drift. The parish records gave Cavalli-Sforza a way to test the idea. They showed that most people in the mountain valleys high above Parma married within their own small villages. Genetic drift is more obvious in such small interbreeding populations because an individual who has many children can flood a population with distinctive genetic variants. In contrast, on the plains around the university, the genetic effects of any one person are reduced, because the interbreeding population is larger. There genetic drift should be less obvious.

One way to measure genetic drift is to look at the relative proportions of various blood groups — such as the A, B, O, and Rh designations on our blood-donor cards — within certain communities. So Cavalli-Sforza and a few assistants took needles and test tubes and fanned out over the countryside. With the help of parish priests they gathered blood samples, often in church sacristies after Sunday mass. They found that the distributions of blood groups varied much more from village to village in the mountains than they did in the towns in the valley, just as predicted by the theory of genetic drift.

His success with the parish records led Cavalli-Sforza to consider the problem more broadly. If he could link blood groups with mating and migrations among the people around Parma, why couldn't he do the same thing on a larger scale? In fact, he ought to be able to determine the genetic relationship between *any* two groups of people by comparing their distinctive biological markers.

More kinds of blood groups, in addition to the A, B, and O types, were being discovered in the 1950s. By 1961 Cavalli-Sforza decided that he had enough data to try his idea. He and a colleague analyzed the blood-group data from fifteen populations — three each from Europe, Africa, Asia, and the Americas and one each from Australia, New Guinea, and New Zealand. They put the results into the first computer Cavalli-Sforza ever used and produced a tree showing how the various groups were related.

The results looked reasonable. The Native American populations from Arizona and Venezuela were closely related to the Eskimos and Koreans in the sample, squaring with the idea that their ancestors had migrated from Asia across what is now the Bering Strait. The Africans and Europeans were genetically close, reflecting their continents' relative proximity. But the work stalled at this point. The blood-group data were spent. Without other genetic information, nothing more could be said about the human family tree.

The best way to determine the genetic relationships among people is to know the exact sequence of the nucleotides in their DNA. But in the early 1960s, when Cavalli-Sforza constructed his first tree of the links among peoples, those sequences were inaccessible. Manipulating DNA in the laboratory at that time was like playing the piano with a baseball bat — the existing tools were far too awkward to examine individual nucleotides. Cavalli-Sforza turned to the next best thing: the thousands of proteins in the human body. The sequence of nucleotides in DNA dictates the sequence of amino acids that constitute proteins, although the translation from DNA to proteins is a convoluted process that partly obscures the underlying nucleotide sequence. Still, by studying protein sequences, Cavalli-Sforza was able to learn at least a little about the DNA differences among people.

Almost immediately he began to find striking patterns of genetic variation. The first region Cavalli-Sforza studied was Europe, where the protein data were most complete. There he saw a prominent wave of genetic change emanating from the Middle East. Some genetic variants occurred at a relatively high frequency in southeastern Europe and then declined in frequency toward the northwest. Other variants showed the opposite pattern; they were more common in England and western France than in the Balkans and Greece. Cavalli-Sforza asked himself what could account for these smoothly varying genetic gradations.

Working with the archaeologist Albert Ammerman, he developed a hypothesis. As people began farming in the Middle East 10,000 years ago, their populations grew. Eventually the increasing numbers of mouths to feed would put pressure on available resources. At that point the sons and daughters of farming families would likely leave their parents' farms and move into new territory. There they either interbred with the existing hunter-gatherers or drove them away.

Cavalli-Sforza and Ammerman proposed that this process accounted for the principal pattern of genetic variation seen in Europe. By the time agriculture was invented 10,000 years ago, the genetic patterns in Europe and the Middle East were somewhat different, since the two populations had been partly separated for thousands of years. Therefore, as farmers spread into Europe from the Middle East, they brought a set of genetic variants different from those in Europe. These were superimposed on the European variants as farmers interbred with the existing populations. The new variants were most common in southeastern Europe, where the farmers entered the continent in greatest numbers, and least common in the northwest, where the farmers' DNA was most diluted by intermarriage. Only in areas unattractive to farmers, such as the Pyrenees, did the genetic variants of the original Europeans stay comparatively intact.

Reaction to Cavalli-Sforza's hypothesis came quickly — and was sharply negative. The idea was based on "untenable assumptions" and "erroneous interpretations," archaeologists said. In the early 1970s, when Cavalli-Sforza and Ammerman first proposed it, archaeologists were almost universally opposed to explanations involving large movements of people. For one thing, migrations are very hard to document in the archaeological record. If a new technology suddenly appeared in a region, archaeologists preferred to explain the change by saying that the existing people had learned a new way of doing things. Political considerations were also a factor. Mass migrations smacked of abhorrent political ideas — such as the notion that the Germans were somehow descended from a tribe of Aryan supermen who swept into Europe from the Asiatic steppes. In the decade after the 1960s, with its emphasis on overcoming animosities between human groups, the idea of an agricultural people streaming out of the Middle East and overpowering the indigenous people of Europe was automatically suspect.

One of the things I love about science is that the questions it asks are, at least in principle, answerable. If a question cannot be answered on the basis of empirically gathered evidence, then that question does not fall within the realm of science. Some questions may remain in limbo for many years before the necessary evidence can be gathered. Some questions may never be answered because the necessary evidence cannot be obtained. But even the discovery that the evidence is not available is a contribution, if only because it allows scientists to move on to new questions.

In the thirty years since Cavalli-Sforza and Ammerman made their proposal, archaeologists have taken a much closer look at the spread of agriculture into Europe. What they have discovered is a process much more complex than was previously recognized. The spread of agriculture involved hunter-gatherers as well as farmers interacting in complicated ways and has resulted in equally complicated patterns of genetic variation.

According to current understanding, farming in Europe appeared

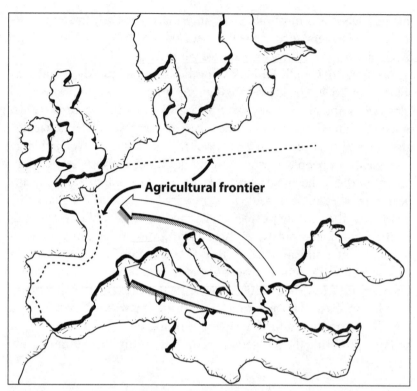

Agriculture moved into Greece from the Middle East about 9,000 years ago, then spread in two great arcs into the rest of Europe: north through the central river valleys and west along the Mediterranean coastline. By about 7,000 years ago a clear boundary had taken shape in western Europe. To the east were farmers and scattered bands of hunter-gatherers. To the west were hunter-gatherers who, as they began to experiment with agriculture, also constructed immense stone megaliths.

in what is now Greece about 9,000 years ago. Outside a cave not far from Athens, people were growing wheat and barley and herding cattle, sheep, and pigs. These farmers may have domesticated some of their crops and animals from local strains, but others were clearly imports from the Middle East. From that beachhead, farming expanded into Europe in two directions. Groups of farmers — or at least the idea of farming — traveled west along the Mediterranean shoreline. By 7,000 years ago, farming communities had sprung up on the Italian peninsula, Sicily, Crete, and even along the coasts of France and Spain.

An even more dramatic expansion took place north of Greece.

From the floodplains of rivers in the Balkans, farming communities spread northwestward along the great river systems of central Europe. By 7,000 years ago agriculture had taken hold in a vast arc stretching from Greece to central Germany. The first farmers of central Europe raised their crops on loess, the easily worked soil that had been ground up by the northern glaciers and blown onto the central European plains. These people often lived on the terraces above rivers, where they had access to both water and level ground for planting. Their culture gets its name from a characteristic kind of pottery they made incised with lines and dots — the Linear Pottery culture. They clearly coexisted with remaining populations of hunter-gatherers, who tended to occupy areas less suited to agriculture. Yet the appearance of the farmers is often so abrupt, and their culture so different from that of their foraging neighbors, that it seems plausible that they came from elsewhere. At least in major portions of southern and central Europe, agriculture really does seem to have resulted from migrating people, not migrating ideas.

Somewhere between 6,000 and 7,000 years ago, the Linear Pottery culture had spread into eastern France and northern Germany. But then its expansion halted. For nearly a thousand years a clear frontier existed. On one side were the farmers of central Europe, interspersed with a dwindling hunter-gatherer population. To the west were populations consisting entirely of foragers.

Groups on either side of the frontier undoubtedly interacted. Gradually the hunter-gatherers began to use clay vessels, and they may have raised sheep. They also began to build the earliest monuments (although the Stonehenge we see today dates from a later period).

The traditional theory about the monuments of northwestern Europe is that they were a by-product of agricultural surpluses. As hunter-gatherers adopted farming, so the theory went, they produced an abundance of food and large populations that could devote their time to huge public-works projects.

But the more archaeologists have looked at this picture, the shakier it appears. Relatively few signs of agriculture date to the period when the earliest monuments were being built. In fact, little evidence for permanent structures has been found from this period. The people who built the earliest monuments may well have been relatively mobile bands of hunter-gatherers, just like their ancestors. But in one respect they were not like their ancestors, because those earlier people had never built these vast stone edifices. Somehow the approach of farming was triggering a critical social transformation.

The hunter-gatherers of northwestern Europe seem to have been working out something in their minds. Before the advent of agriculture, people had been part of the natural world, living in equilibrium with the animals they hunted and with the plants they gathered. They trusted that the world would provide for them so long as they lived on its terms.

The monuments seem to signal a break with this worldview. By building structures, the hunter-gatherers were changing their environment in permanent ways. They were establishing their dominion over the land, separating themselves from nature by claiming certain areas for human use. I hesitate to use the term religion here, because anthropologists do not know anything about the beliefs of these people. But they may have been undergoing something like a religious conversion.

Not everyone agrees with this picture, of course. One obvious question is why the same monument-building process did not occur elsewhere in the world. Still, the idea has a certain intuitive appeal. The hunter-gatherers of northwestern Europe needed to change how they thought about the world before they could commit to agriculture. Monumental architecture created the attitudes that made farming acceptable to these people. It provided a sense of place, a bond with a particular location, that allowed them to use that location in new ways.

Following the first wave of monument building, agriculture swept across the rest of the continent, faltering only in the frozen north (which is still occupied by pastoralists and hunters). Within a couple of millennia, the old hunting-gathering culture was gone almost everywhere in Europe, especially as domesticated herds encroached upon the lands of wild animals. Over time, as agriculture became established, the monuments fell into disuse. Places like Stonehenge remained centers of ritual and mystery, but even that purpose gradually became less compelling. By the time of the Roman invasions, the rationale for the monuments had been forgotten.

Today people gather at Stonehenge and at the other monuments of northwestern Europe to try to recapture the worldview of people who lived long ago. The irony is that the monument builders may have been looking in the opposite direction. They may have been trying to accommodate themselves to the coming of the modern world.

If descendants of Middle Eastern farmers settled in some parts of Europe while hunter-gatherers remained in others, one might conclude that some Europeans are descended from the first group and others from the second. But such a conclusion would be mistaken, because it overlooks the inevitable genetic mixing that occurs among groups. Over the millennia people in Europe mixed so thoroughly that everyone of European descent now has some DNA from a Stone Age hunter-gatherer and some from a Middle Eastern farmer. This is a relatively recent conclusion, and the best way to explain it is to return to the scientific odyssey of Luca Cavalli-Sforza.

In 1968, while he was working on the research that would lead to his proposals about the peopling of Europe, Cavalli-Sforza made a migration of his own. The geneticist Joshua Lederberg, with whom Cavalli-Sforza had done his early research on bacterial genetics, had become chairman of the genetics department at Stanford University; he invited Cavalli-Sforza to come to California and do research for a year. It was a tumultuous time in the United States. The area around Stanford was brimming with antiwar activity, and Cavalli-Sforza's children were soon marching the streets to protest the Vietnam War. But Cavalli-Sforza liked the intellectual vitality of Stanford, and in 1971 he moved to California for good.

Unbeknownst to Cavalli-Sforza, a series of experiments then going on in the Bay Area was about to transform the study of genetics — and much of the rest of biology. A pair of biochemists from Stanford and the University of California at San Francisco, Stanley Cohen and Herbert Boyer, were figuring out how to cut DNA at precise locations, combine DNA from different organisms, and grow the resulting hybrid DNA in bacteria. For the first time, human beings could control the exact nucleotide sequence of DNA. The age of genetic engineering had begun.

Now Cavalli-Sforza finally had the tools he needed to look at DNA directly, not just at how those differences are expressed in proteins. Soon he and his colleagues were working on a wide range of projects connecting genetics with history. Cavalli-Sforza was involved in a study of mitochondrial DNA at the same time that Allan Wilson and his students were doing their work on mitochondrial Eve; he and his collaborators at Stanford would have announced roughly similar results within a few months if Wilson had not published the findings first. Meanwhile other students, postdocs, and research associates were looking at other parts of the genome. Today many people around the world who are using genetics to study history have spent some time in Cavalli-Sforza's lab.

Some of the most interesting work done at Stanford has been on the Y chromosome. For many years geneticists thought that the Y chromosome would not be of much use in studies of human history. Though it is much longer than mitochondrial DNA — 60 million bases compared with about 16,500 — the Y seemed to have relatively few genetic variants. Then two researchers in Cavalli-Sforza's lab — Peter Underhill and Peter Oefner — developed a new way of finding variants on the Y. Within a couple of years they had detected more than two hundred locations on the Y chromosome that differ among individual males. What's more, these variants turned out to be almost ideal for tracking the movements of males across the globe, because different variants are tightly clustered in particular geographic regions. "I'm obviously biased," says Underhill, "but if you had to choose just one genetic system to track human migrations, it would have to be the Y, because it has the largest number of unique genetic markers and is a sensitive index of population bottlenecks and founder effects."

The peopling of Europe has been a particular focus of Underhill's work. He and his colleagues have identified specific mutations of the Y chromosome that appear to be associated with historical periods. For example, they have identified several Y-chromosome haplotypes that were apparently carried into Europe by farmers from the Middle East. These haplotypes are especially common along an arc extending from Greece into central Europe and along the coastlines of the Mediterranean. If the haplotypes are taken to represent the genetic contributions of Middle Eastern farmers, then about 22 percent of today's European males received their Y chromosomes from this source.

Studies of European mitochondria have produced very similar results. At the University of Huddersfield in England, Martin Richards is one of a group of researchers who have applied to Europe a new way of tracing mitochondrial haplotypes. First, Richards and his colleagues analyzed the mitochondrial DNA of the two groups likely to have been source populations for all of today's Europeans — the people of western Asia (including the Middle East) and those of northern Africa. Then they compared the mitochondrial haplotypes of the descendant populations with those of the source populations to determine how long they had been separated. Corrections had to be made to account for continued genetic exchanges in both directions between the descendant and source populations. But DNA is such a rich source of historical information that these corrections are generally possible, so long as the number of mitochondrial sequences being analyzed is large enough.

In a study of more than 4,000 people from Europe and western Asia, Richards and a team of other geneticists found that the mitochondrial DNA now in Europe reflects several broad waves of immigration into the continent. The first wave, representing about 10 percent of European mitochondrial DNA, dates to the initial colonizations of Europe by modern humans. The majority of the mitochondrial DNA in Europe arrived during the waxing and waning of the Ice Age, probably as the result of a continuous trickle of people from the Middle East into more northern climates. Finally, about one-fifth of the mitochondrial DNA appears to date from the movement of Middle Eastern farmers into Europe — about the same percentage as that calculated from the Y chromosome. "A consensus is emerging on what the genetic data are telling us," Richards says. Underhill agrees: "It's remarkable that the numbers are so similar."

Indeed, a four-to-one split for the genetic contributions of pre-agricultural Europeans and of migrating Middle Eastern farmers is a figure almost everyone accepts. Cavalli-Sforza points out that he and Ammerman never insisted that all the DNA in Europe was from Middle Eastern farmers — their calculations attributed only about 28 percent of the genetic variation to this source. And the archaeologists who disputed Cavalli-Sforza and Ammerman say they never claimed that the Middle East and Europe were separated by an impermeable barrier; they simply wanted to emphasize the cultural and genetic contributions of Europe's preagricultural inhabitants.

If particular mitochondrial DNA and Y-chromosome haplotypes are associated with the spread of farming into Europe, then every European can trace his or her mitochondrial DNA and, in the case of males, Y chromosome either to a preagricultural hunter-gatherer or to a Middle Eastern farmer. The existence of distinct categories of haplotypes might seem to imply that the two groups have remained somewhat separate. But they haven't, as demonstrated by the distribution of the haplotypes. Every European population studied has had haplotypes that trace back both to early hunter-gatherers and to farmers. This could have happened only if the descendants of these two groups had extensively intermarried. This intermarriage would have thoroughly scrambled the DNA in the chromosomes of each person. Particular haplotypes are more common in some places than in others — that's how human migrations can be tracked. But in no case has a European population been found to be descended entirely from a single group. Even the Basques, long cited as a bastion of preagricultural DNA, have plenty of mitochondrial DNA and Y chromosomes from Middle Eastern farmers.

The next chapter describes the historical processes that led to this extensive degree of genetic mixing in Europe. The important observation is that such mixing is the rule, not the exception. Whenever groups of modern humans have come into contact they have begun to blend their DNA. And in the case of Europe, this mixing probably will be even more extensive in the future than it has been in the past.

10

Immigration and the Future
of Europe

Liberté, Égalité, Fraternité.

— French revolutionary slogan, 1789

A GOOD PLACE to glimpse Europe's multiethnic future is the outdoor market held every Sunday morning in Villeurbanne, a suburb on the east side of Lyon, France. Alongside trim apartment buildings and well-tended parks, the sidewalks and streets are thronged with people from everywhere in the world. Behind makeshift wooden tables, vendors from Senegal, Israel, Slovakia, Pakistan, and China hawk shoes, leather coats, perfume, and scarves. The buying public is slightly less diverse, but not much. Immigrants and their children now account for about 9 million of France's 60 million people. The percentage is even higher in a city such as Lyon — the country's second largest city and a magnet for immigrants because of its location midway between Paris and the Mediterranean.

The percentages of immigrants in France and other European countries are not yet as high as in the United States. But immigration has become a controversial issue in Europe in recent years. A couple of generations ago the majority of immigrants to France were from other parts of Europe, so they had the same broad European culture, and their physical characteristics were much like those whom the French call *français de souche,* or root French. In the last few decades, however, the majority of French immigrants have come from the Maghreb — the North African countries of Morocco, Algeria, and Tunisia — as well as from sub-Saharan Africa, the Middle East, and Asia. The result has been a tumultuous mixing of cultures, languages, and appearances. In a place like the market of Villeurbanne, where everyone speaks French but with accents that would make the minister of culture cringe, the term *lingua franca* once again seems appropriate.

Today Villeurbanne is an orderly, clean, and busy place. Yet just five years ago this suburb and its neighbors were the scene of several days of looting and rioting. A young Algerian-born immigrant named Khaled Kelkal, who had been involved in an unsuccessful attempt to blow up a high-speed train, was killed by uniformed security forces, and his dead body was shown on national television. In response, gangs of Maghrebi youth went on a rampage. They burned cars, attacked shops, and smashed telephone booths and bus shelters. Instantly the suburbs of Lyon became a symbol for the single most visible division in French society: that between the comfortably employed and long-established middle class and the nation's economically disenfranchised, politically embittered immigrants.

The municipal and national governments of France seem less willing than governments in other countries to write off entire sections of their cities. In recent years the suburbs on Lyon's east side have received a new light-rail line, improved housing, and new schools. At the same time, a surge of economic growth has taken the edge off ethnic tensions. "If you really want to work in this country, you can succeed," says a waiter at a nearby café, whose family moved to France from Turkey.

Still, I was warned against going to certain places in Lyon, especially around the high-rises that border the inner city. And for all the economic progress of the 1990s, one gets the sense that an economic downturn could redouble ethnic antagonisms. "In France there has always been a lot of racism against 'Arabian people,'" says Djida Tazdait, a leader of the Jeunes Arabes de Lyons et Banlieue (Young Arabs of Lyon and the Suburbs). "It's hard for people with our ethnic origin for a lot of reasons — because of the visibility of our physical differences and our religion. We are discriminated against in politics, administration, companies, and so on. The new generation has identified the problems, and we know that it's necessary to fight them. But how many generations will be necessary?"

Like immigrants around the world, Tazdait has experienced first-hand the prejudice often encountered by newcomers. She was six when she and her family moved to France. "We came from Algeria right after the Independence War," she says. "We had to leave because our country was very poor. People worked hard to rebuild after the war, but they were not paid, and after a few months it became impossible to stay. Also, France needed a lot of immigrants, so a big wave of immigrants came here thinking they would find 'El Dorado.'

"We lived in a village, in an area of Lyon near a garbage dump. The house was big but not very comfortable. It had no warm water, and the toilets were outside. But it had a garden, which was very important for us. We cultivated it very hard to produce food. It was important to us, because it reminded us of our Algerian life.

"My parents wanted very much for their children to succeed. I was the oldest of five children, so I had to be a good example for my sisters and brother. I went to high school, college, and medical school. But we were not very numerous at this level in the universities. And I encountered racism with the police every time I went out at night. They controlled our identity with tommy guns, speaking with lots of racial insults, and using *tu* instead of *vous*. I understand now why young people hate the police, because those controls are very degrading. But at that time we thought it was normal, because our parents were resigned to the police, too."

Controversy over immigration has cast a pall over a country that remains committed to the revolutionary ideals of liberty, equality, and brotherhood. As in other European countries, right-wing politicians have sought to capitalize on concerns over immigration. During the mid-1990s, the far-right National Front won control of three large French cities, including Toulon, the country's largest Mediterranean port. The party ran on an anticrime, antiunemployment platform. But in its call for "French-first" policies that would discriminate against immigrants and their families, the National Front did not disguise its core agenda. "In France the far right often speaks about crime or security," says Florence Veyrié, who works for Citizens Forum in Lyon, a voter education group. "But they are really talking about immigration."

The National Front has never controlled more than 15 percent of the French vote, and the economic growth of the late 1990s further diminished its influence. Today the official position of the French government remains one of assimilation. The children of immigrants are expected to absorb the language and culture of their new country so that ethnic and national origins will be irrelevant by the second generation. France is so committed to integration that it doesn't even collect information about the different groups within its population. It believes that affirmative action programs such as those in the United States are forms of "positive discrimination" that reinforce invidious distinctions.

Nevertheless, the recent success of the National Front reflects a more deep-seated ambivalence in French life. In polls more than half the French say that too many Muslims are in the country. Many doubt that immigrants from non-European countries can be successfully integrated into French society. They worry that the immigrants of the last few decades will become a permanent minority, undermining France's egalitarian ideals.

I focus on France in this chapter because it displays so many of the crosscurrents immigration has generated in Europe. But every European country has a different relationship with immigration, and using France as an example of European attitudes in general can be misleading. Spain and Italy have fewer immigrants than does France, yet anti-immigrant sentiment has been much higher in those countries. Germany is less committed to assimilation, though the growth in the number of foreigners living there — from 4 million a decade ago to 7 million today — has sparked a furious debate about immigration. The countries that border the Mediterranean or Asia, which must deal with boatloads and truckloads of immigrants trying to slip across borders, face different issues than do countries in the interior.

Despite their differing concerns, immigration is near the top of the political agenda for every European country. As birthrates continue to fall, many nations will have to admit many more immigrants to keep their economies from stagnating. Because of the increasing openness of borders within the European Union, immigrants can travel easily from one country to another once they get to Europe; in essence, the immigration policy of the least restrictive nation could become the policy of all. And illegal immigration has become a major problem throughout Europe, especially as undocumented workers are shunted into sweatshop work, crime, and prostitution.

Europeans have discovered that their future is indissolubly linked with immigration. Yet that is nothing new. The history of the continent is inseparable from considerations of immigration — within the continent, from other parts of the world into Europe, and from Europe to the rest of the world. The human history of Europe is thoroughly tangled — genetically as well as culturally — and today's events have many parallels with what has happened in the past.

At the same time, immigration in Europe is raising new and unprecedented issues. Groups are coming into contact on a scale and with an intensity that has rarely occurred before. Economic, social, and political forces are minimizing some differences between groups while accentuating others. And Europe's history, which repeatedly has emphasized the differences rather than the commonalities among people, inevitably complicates the decisions that must be made.

Past efforts by European and American scholars to draw rigid distinctions among human groups would be comic if they weren't so tragic. Scientists have devoted entire lifetimes to cataloging and measuring how human beings differ. They have compared skin colors, pored over the nooks and crannies of skulls, and measured the lengths of arms, legs, torsos, and genitals. For the most part this endless poking and prodding has been pointless. Scientists today know little more than what a cursory examination of the world's people reveals — that individuals differ physically in ways that correspond roughly but not perfectly with their ancestry. Nevertheless, the hypotheses developed in the course of all this speculation about human differences have profoundly affected history. Ideas that we now know to be utterly false have had an indelible effect on the world.

One of these ideas is that human beings can be divided into discrete races with specific physical and often cultural characteristics. In some parts of the world this idea still seems self-evident, and most people probably assume it has been a commonplace of all human societies. But many modern scholars disagree. They contend that Europeans invented racism in the eighteenth and nineteenth centuries to distinguish between European civilization and the other peoples and cultures of the world. These scholars point out that ancient civilizations rarely associated physical appearance with character. Intergroup marriages were common in the Hebrew Bible; Moses himself married an African woman. The Greeks and Romans carefully observed the different appearances of people, but they attributed these differences to the environment rather than to inner attributes — for example, they thought that Ethiopians had black skin because they had been burned by the heat of the sun. The crucial distinction for the Greeks and Romans was between civilization and barbarism, and even the most exotic-looking barbarian could become civilized by adopting Greek or Roman culture.

Other sociologists and historians reject the notion that racism is a recent invention. They observe that people have a natural tendency to divide the world into "us" and "them." If differences in physical appearance can be used to reinforce this division, they will be. If the distinction is entirely cultural, well, appearances can be deceiving. Some scholars make the case that this tendency to divide the world into groups is biological. They suggest that evolution selected for what has been termed "ethnocentrism" because this behavior contributed to a group solidarity that helped people survive. Personally, I suspect that prejudice against others is mostly a learned response, although a very powerful one. To survive, people have to live in societies, which are inevitably structured in terms of relationships among individuals and groups. From an early age, children learn that they should trust certain individuals — family members, for instance — and be suspicious of others. That this lesson should endure throughout life is hardly surprising.

Even if European scholars of the eighteenth and nineteenth centuries did not invent racism, they gave it a theoretical backing and practical significance that it had never had before. Other books tell the story of the development of racism in Europe and America, and I won't repeat the tale here. But I would make a couple of important observations. First, European racism has always been entangled, in both positive and negative ways, with Christianity. On the positive side, the Bible proclaims that all people descend from a single couple — Adam and Eve. According to the church scholars who studied the Bible's genealogies, Adam and Eve did not live that long ago — just a few thousand years before the birth of Christ — and that account was often used to argue for the unity of mankind.

But Christianity, like most religions, has often done as much to justify prevailing inequities as to challenge them. For example, Christianity profoundly influenced European ideas about the Great Chain of Being, a quasi-scientific theory arising from both religion and classical science that ranked living things, including human groups, into a hierarchy of moral worth. These ideas led many Europeans to conclude that they were higher on the chain than were other groups, closer to angels than to beasts.

Faith in biblical authority began to break down in the early 1800s.

Scholars pointed out that the differences among groups were too great to have emerged in just a few thousand years. Either some humans must have been living elsewhere in the world at the same time as Adam and Eve — or the world was considerably older than the Bible said it was. In either case, it seemed eminently reasonable that different groups could be the products of separate acts of creation. These "polygenic" theories became increasingly popular in the nineteenth century. In the Dred Scott decision of 1857, which helped spark the American Civil War, Chief Justice Roger Taney wrote that the treatment of Africans in America was justified because they were "beings of an inferior order."

The 1859 publication of Darwin's *On the Origin of Species* simply moved speculation about the origins of human groups from the realm of religion into that of biology. Though Darwin himself argued against polygenic ideas, others quickly invented evolutionary reasons for the differences among groups. European biologists, for example, immediately assumed that the people of Europe had somehow progressed higher on an evolutionary scale than had other peoples. Some went even further, positing that different groups had evolved from different species of apes. Ernst Haeckel, an anatomist who was the leading advocate of Darwinism in Germany, wrote in his 1905 book *The Wonder of Life*: "The lower races — such as the Veddahs [aboriginal Sri Lankans] or Australian Negroes — are physiologically nearer to the mammals, apes and dogs, than to the civilized European. We must, therefore, assign totally different value to their life."

If different groups were the products of different evolutionary trajectories, then they should differ physically as well — after all, that's how naturalists had always distinguished among species. In the second half of the nineteenth century, the desire to find such differences fueled a mania for human measurement. These measurements focused especially on the skull, since the brain seemed the logical source of differences in behavior and thinking. Sure enough, after measuring the volumes of thousands of skulls, race scientists concluded that Caucasian skulls were on average the biggest.

These measurements have long since been thoroughly discredited. They resulted from picking the biggest skulls from Caucasians and smaller skulls from other groups, along with other obvious biases. But even the original skull measurers were not entirely satisfied with their results, for they found that skull volumes within each group fell along a continuum, and these continua overlapped among groups. Some Africans had larger heads than some Caucasians, and vice versa. Ranges that overlapped so extensively provided no clear way to distinguish between groups.

Another popular nineteenth-century measurement involved something called the "facial angle," which is essentially the angle formed by the base of the skull, the bottom half of the face, and the top half of the face. A relatively large facial angle — close to 90 degrees — was supposed to denote biological superiority, while a smaller angle was an indication of evolutionary backwardness. But the results of this measurement were as muddled as those of skull volume. While the average facial angle of one group might differ from the average of another group, the results weren't consistent between individuals — as anyone can see by observing people on the street. As with brain size, the overlap between groups ruled out the use of the facial angle as a way to sort people into categories.

A final head measure that achieved great popularity was called the cranial index, which compared the maximum width of a person's skull to its maximum length. People with relatively narrow heads have cranial indexes of 0.75 or less. People with relatively wide heads have cranial indexes exceeding 0.8. According to the nineteenth-century skull measurers, narrow heads were a mark of intelligence and culture, whereas broad-headed people belonged to an obviously inferior "race."

In the early decades of the twentieth century, this index also was completely discredited. It became decidedly less popular when Africans and Australian aborigines were found to have some of the narrowest heads. Also, the cranial index turned out to be much less fixed than had been thought. A child with a narrow head could grow into an adult with a broad head. When broad-headed immigrants came to the United States, many had narrow-headed children.

But before this notion was discredited, the cranial index played a prominent role in some of the most bizarre racial theories ever concocted. According to the most notorious of these, the people of Europe could be divided into three distinct races. The Nordics of northern and northwestern Europe were tall and fair and had narrow heads. The Mediterraneans of southern Europe were short, dark, and also narrow-headed. In between were the Alpines, who were short, brown-haired, and broad-headed. These categories have only the vaguest relationship to the characteristics of actual Europeans; in fact, people with all of these features can be found in every part of the continent. Yet for a time these ideas were used to divide the people of Europe into racial categories just as rigid — and as damning — as the racial categories applied to the rest of the world.

In Germany this theory took a somewhat different form in the late nineteenth and early twentieth centuries. There the Nordics became the Aryans, a warlike Indo-European people who had supposedly swept into northern Europe from the Asian steppes, displacing the inferior broad-headed people then living there. The popularity of this theory marked the beginning of the myth of the blond, blue-eyed, superior Aryan. Before then, the blue-eyed blond was usually associated with a different stereotype, that of the dreamy romantic.

All such theorizing could be dismissed as harmless academic sophistry if not for the political potency of these ideas. In 1921, in response to a growing interest in eugenics in Germany, a Munich publisher brought out a book called *Outline of Human Genetics and Racial Hygiene*. One of the authors was the geneticist Fritz Lenz, a leading advocate of the Aryan ideology. Another was Eugen Fischer, an anthropologist who had studied and condemned the mingling of Europeans and Africans in Southwest Africa. The third was Erwin Baur, a botanist who specialized in genetics. The publisher gave a copy of the 1923 second edition of the book to Adolf Hitler, who read it carefully while in prison following the Beer Hall Putsch. The ideas of the book were obviously prominent in Hitler's mind as he wrote *Mein Kampf*.

In retrospect, all of these ideas about the existence and origins of European "races" are so ludicrous that it is shocking to realize how much effort went into formulating, defending, and debunking them. Over the course of the twentieth century, historians, archaeologists, and geneticists learned much more about the movements of people during European history and prehistory. These movements bear no resemblance to the fantasies of the Aryan supremacists. What they resemble is something quite different: a tangled mass of yarn, thoroughly snarled yet all of one piece.

The bulk of Europe remained a prehistoric society — that is, a society without formal writing — until it came into contact with the Romans. But even as the curtain of history rose on the continent, various groups were already wheeling about the European stage. Rome's first contact with the rest of Europe came during the great Celtic expansions of the fifth through the third century B.C. The Celts were a people who, judging from the archaeological record, loved both art and fighting. They filled their graves with beautiful gold ornaments and statues — right next to the swords and shields they carried into battle. As one contemporary observer put it, "The whole race is madly fond of war, high-spirited, and quick to battle, but otherwise straightforward and not of evil character."

Actually, determining whether the term "Celtic" should be applied to a people or to a culture is not easy. Massive population movements certainly occurred within Europe from 500 to 200 B.C. At the beginning of this period, groups speaking Celtic languages moved from central Europe into northern Italy, Hungary, Transylvania, and Macedonia. Celts also established themselves in the far northwestern corner of the continent, including the British Isles, which is the only place where remnants of their languages survive today. But their culture undoubtedly spread more widely than their DNA, as other Europeans adopted a way of life that met with great success for a time.

The Celts were just one of many highly mobile barbarian tribes described by classical writers. Today the names of those tribes are little more than dusty historical relics: Alamans, Angles, Saxons, Jutes, Lombards, Vandals, Gepids, and Alans; Goths, Franks, Slavs, and Teutons; the Huns, Avars, Magyars, and Mongols. Yet in their day those names struck terror into the hearts of the inhabitants of civilized Greece and Rome. Later European historians would look back at these loose agglomerations of people and see the first faint stirrings of nation-states. Such interpretations are mostly wishful thinking. Europe was undergoing great instability and population movements during the time of the Roman Empire. Groups coalesced for a few generations, dispersed, and reformed under new names. Tribes undoubtedly blended into one another, with a constant shifting of alliances, clans, and mates. This genetic churning of history is an important way in which the mitochondrial haplotypes and Y chromosomes of Europeans became so mixed.

Nor did the churning stop at the borders of Europe. In geopolitical terms, Europe was simply the western part of Eurasia in Roman times, with no obvious dividing line between East and West. Events taking place half a world away could have consequences on European soil. For example, in the first century B.C. the Han dynasty in China destroyed the empire of the Xiongnu in what is now western China. The defeated people, with the herds of cattle that supported their populations, began pushing westward. By the second century A.D. these people, or the populations they had displaced, were living north of the Caspian Sea, where they were known as the Huns. By the fourth century the Huns were in the vicinity of the Ukraine. There they clashed with a Germanic tribe known as the Ostrogoths, pushing the Ostrogoths and the neighboring Visigoths into Rome. In 443 the Huns came under the command of their greatest leader: Attila. He and his warrior armies moved into the heart of Europe, forming coalitions with various European tribes and fighting others along the way. The Huns and their allies swept across Germany and eastern France but spared Paris — traditionally thought to have been saved by the prayers of Saint Genevieve. Finally, near Troyes, a coalition of Ostrogoths and Franks turned back Attila and his armies.

The Huns left more than dead Europeans in their wake. A French friend of mine, a strikingly beautiful woman, is from the town of Troyes. When we were talking about family histories, she pointed to her eyes and noted that they are unusually slanted — what the French call *yeux bridés*. It turns out that her family has always traced its descent in part from the Huns. She also told me that all three of her children were born with what is called a Mongoloid spot — a bluish patch near the base of the spine that gradually fades with age. Though not unknown outside Asia, such a mark is much more common among people of Asian ancestry.

The populations of Europe and Asia were not only mixing, they also were growing. Before the advent of agriculture, the population of Europe was probably less than 1 million. By the first century A.D., it had expanded to approximately 30 million. However, this population growth was very uneven. Some groups and regions saw a much greater expansion, particularly in areas well suited to agriculture, than did others. In turn, expansions contributed to migrations from one area to others, further mixing the European gene pool.

The history of Europe is too convoluted to trace the movement of DNA across the entire continent. But France offers a good example of the process at work. Today the French see themselves as a single people. Yet they are an amalgamation of a huge number of ancestral groups. At the time of the Celts, no fewer than seventy-four tribes

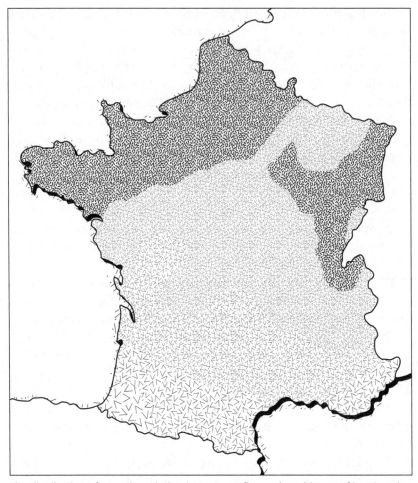

The distribution of genetic variation in France reflects a long history of immigration and population movements. The variants in the northeast probably record the movement of Germanic people, including the Franks, into Roman Gaul. The variants in the northwest likely represent the waves of settlers from England and Scandinavia from the fifth through the tenth century. The Basques in the southwest and Mediterranean people in the southeast also have influenced genetic patterns. The distributions shown here account for about 18 percent of the total genetic variation in France.

lived in modern-day France, most occupying tracts of cleared land separated from other open areas by forests. On these tribal groups were superimposed other populations who came later to France. Romans began to settle in southern France in the second century B.C., gradually extending their control to the north. The Franks, from whom the country would get its name, were a Germanic people who crossed the Rhine and gradually pressed southward. In the fifth and sixth centuries, Celtic refugees from the Anglo-Saxon invasions of the British Isles settled in northeastern France, followed by Viking settlers in the ninth and tenth centuries. There were also the Basques in the southwestern corner of the country, the Burgundians, who had migrated to southeastern France from the southeastern Baltic, the Saracens from northern Africa, the Magyars from the east, and others.

Waves of immigration have continued into modern times. Yet, remarkably enough, some of the genetic patterns established during the early historical period in France can still be detected in the DNA of the French. By analyzing the differences in proteins among people from various parts of the country, geneticists have drawn a map showing components of genetic variation. The distinctive regions in the northeast and northwest of this map probably reflect the movements of Franks, Celts, and Vikings. The Basque influence, in the southwestern corner of the country, can be seen on the map, as can a genetic signal in the southeast that is probably related to the movements of various Mediterranean people.

During the Middle Ages the population of France grew largely from within. By 1300 the population within today's borders was about 20 million, up from around 5 million in Roman times. The population stagnated at the end of the Middle Ages — partly because of the loss of a third of the people to the Black Death in the fourteenth century, with continuing losses from disease in succeeding centuries. But by the beginning of the industrial age in 1800, the population had recovered and stood at perhaps 30 million.

Suddenly France's birthrates began to fall. France was the first country in the world to undergo what is called the demographic transition, from high fertility levels to low. In 1750 the birthrate in France stood at forty live births per thousand people. A century later it was twenty-six per thousand. One factor in this change was a provision in the Napoleonic Code that required peasants to divide their land equally among male heirs, which gave peasants an incentive to keep their families small. Another was the increasing availability of birth control devices, including the condom and the pessary (an early diaphragm). Meanwhile the other countries of Europe, especially Germany, continued to grow. The French soon began to worry that they were losing political power to their more prolific neighbors.

Partly because of this concern, France has been relatively open to immigration for much of its modern history. Over the past two centuries it has absorbed sizable numbers of people from Spain, Portugal, Italy, Russia, Turkey, and other European countries. Most of these immigrants have long since blended into the population. In some cases small immigrant enclaves still exist, especially where they have specialized in trades such as construction.

Immigration to France was particularly high in the 1920s after millions of young French men were killed in World War I. But in absolute numbers the peak came in the 1950s and 1960s, when millions of guest workers moved to France to work in the booming industrial sector. Many were from European countries such as Spain and Portugal, but a large contingent arrived from northern Africa, particularly after France's withdrawal from Africa as a colonial power beginning in the 1950s. By 1974 roughly 4 million foreigners were in the country.

Immigration policies abruptly changed in 1974, when the oil embargo imposed by the Organization of Petroleum Exporting Countries (OPEC) sent the economy into a tailspin. Suddenly many of the immigrants had no jobs. But by this time most were at least partly integrated into French society and had no desire to leave. Though the government curtailed new immigration, many recent immigrants eventually brought their families to France, and political refugees and illegal immigrants continued to find a safe haven there. As a result, the population has continued to diversify.

National governments have a right to determine immigration policies. Yet those policies inevitably are influenced by circumstances outside the control of government. In particular, over the past few decades Europe has been undergoing demographic changes that will be just as consequential as those of the eighteenth and nineteenth centuries. In most European countries birthrates are now so low that the population cannot replace itself. Between 1950 and 1995, France's fertility level plummeted from 2.7 children per woman to 1.7 — well below the replacement level. In other European countries, such as Italy and Russia, the rates of childbearing are even lower. Without continued immigration, the population of Europe will begin to decline in the first few decades of the twenty-first century. (In the United States, by contrast, demographers project that higher fertility and immigration levels will boost the population from 280 million today to 400 million by 2050.)

As Europe's population shrinks, it also will age. France's working-age population will drop from 38 million to 34 million over the next five decades if current trends continue, and its elderly population will almost double — to 15 million. Fewer workers will be supporting more retirees. Because older people spend less and draw on their savings to finance their retirement, investment in the economy will decrease. Without large rises in productivity, Europe's economies seem destined to decline both in relative terms — compared with the rest of the world — and perhaps even in absolute terms as the workforce dwindles.

Many European governments have tried to encourage women to have more children. But the effectiveness of these campaigns has been limited, since women in modern societies have attractive options besides raising large families. European countries therefore have only one way to maintain the size of their populations: accepting more immigrants. For all of Europe (including Russia) to remain as populous as it is today, immigration levels from outside Europe will have to double — to 1.8 million annually. At that level, non-European immigrants and their descendants will account for about 20 percent of the population of Europe in the year 2050 — the highest percentage in its history. In fact, a population infusion of that magnitude into Europe has probably occurred at only one other time in the past 30,000 years: during the spread of agriculture into the continent beginning 9,000 years ago.

New immigrants into Europe will come from many parts of the world, bringing different cultures and physical appearances. It remains to be seen whether these new people will blend in with local populations or whether, at the opposite extreme, parts of Europe will be balkanized into separate ethnic and racial enclaves. Historical precedents exist for both extremes. Viewed over a long enough time period, every European culture is the product of a prodigious mixing of past groups. Yet every European country also has pockets of ethnically distinct groups, such as the Roma (also known as Gypsies), which have remained mostly endogamous for many generations. And as the recent history of the Balkans shows, ethnic divisions always have the potential to sharpen rather than blur.

Nevertheless, the forces of assimilation are very strong. In France, for example, the integration of young immigrants into the mainstream of French society seems only a matter of time. Young immigrants and the children of immigrants mix extensively with nonim-migrants in schools, at clubs, on the streets. They wear the same clothes, listen to the same music, watch the same television shows and movies. (Many of these cultural influences come from the United States, much to the annoyance of the guardians of French culture, but that's another story.) In polls the vast majority of young Muslims say that they consider France their home and want to be a part of French society. Already, 45 percent of Algerian men marry French women.

True, the integration of immigrant youth into the French economy has been slow. But the reasons for that have as much to do with the economic doldrums of the 1970s and 1980s as with cultural differ-ences. As the economy picked up steam during the 1990s, the pace of assimilation seemed to many observers to increase. Sociologist Alec Hargreaves points out in his book *Immigration, "Race," and Ethnicity in Contemporary France* that most of the children of immigrants iden-tify far more closely with the cultural codes of France than with their parents' culture. "The question is no longer whether, but how, immi-grants and their descendants should best be incorporated into French society," he concludes.

One last issue needs to be addressed. Perhaps physical differences will keep people with European ancestors and the recent immigrants from Africa and Asia from mixing completely. I'll return to the issue of intergroup marriages in the final chapter of this book. But the les-son from history is that physical differences do little to slow the mix-ing of groups unless powerful societal forces keep them rigidly sepa-rated. Such separatist sentiments do exist in Europe, as evidenced by the fulminations of various right-wing groups. But these voices are a minority in Europe, and they do not appear capable of gaining politi-cal power, at least in most areas. The blending of peoples that has been going on for millennia in Europe seems destined to continue.

Today almost no one pays any attention to the old "races" of Europe — the Nordics, Alpines, and Mediterraneans who once were thought to embody the achievements of European civilization. In the long run, the physical distinctions between Europeans and the rest of the world's people may seem just as inconsequential.

V

The Americas

11

The Settlement of the Americas

> An Indian is an Indian. If you've seen one, you've seen them all.
>
> — Don Antonio Ulloa, 1772

WHEN I WAS A BOY growing up in the arid cattlelands of eastern Washington State, in the years before camps, computer games, and cable television, when summer vacation stretched in front of us like a long raft trip down a slow river, my friends and I would sometimes make a day of it and ride our bikes to the Tri-Cities, sixty miles away. Weaving between hay trucks and pickups, singing off-color parodies of marching songs that we had learned during Boy Scout conventions, we pedaled past alfalfa and beet fields, potatoes and mint, as the thwack, thwack, thwack of giant sprinklers cast a cool, rainbowed mist against the sun. The first of the Tri-Cities we reached was Pasco, a dusty industrial town of tractor dealerships and auto repair shops. We didn't stop there. We continued on to Kennewick, on the southwestern shore of the Columbia River, a few miles above the river's confluence with the Snake. We would be so hot by the time we reached the shore that we would throw our bikes down on the sand and run shouting into the icy water. I can still see the murky green color of the river as I opened my eyes underwater to soak up that blessed coolness.

If my friends and I had chosen to go swimming a few miles upstream, maybe we would have been the ones to discover the bones protruding from the eroded riverbank. Instead, they were found in 1996 by two teenage boys who were trying to sneak into the annual hydroplane races held on the Columbia River. One of them stepped on something round and hard. When he plucked the object from the mud, he saw that it was a skull.

The boys stashed the skull in a bush so they wouldn't miss the race and later told a security officer what they'd found. By way of the county coroner, the skull and much of the rest of a human skeleton came into the possession of James Chatters, a forensic anthropologist in the Tri-Cities. Initially Chatters thought that the skeleton was that of a white man, a pioneer who had died maybe one hundred years ago, in a fishing accident perhaps. But he noticed a stone projectile lodged in the man's hip, an injury unlikely to have occurred in pioneer days. Chatters decided to send a piece of a finger bone away for radiocarbon dating. The test showed that the bones were approximately 9,500 years old.

Kennewick Man, as he came to be called, was obviously a very early inhabitant of the Americas. No human bones more than about 13,500 years old have been found in either North or South America, and New World fossils as old as Kennewick Man are very rare. Moreover, many of these early American fossils are somewhat distinct from later fossils. Today's Native Americans show a wide range of physical characteristics — the centuries-old stereotype that all Indians look alike is laughably wrong. But many of the earliest occupants of the Americas didn't look much like current Native Americans. Kennewick Man's skull is relatively narrow, with a projecting nose and a pronounced chin. Those traits are more often seen in Europeans than in Native Americans, which is why Chatters originally thought Kennewick Man was a pioneer. Could North America have been settled first by a wayward tribe of Europeans who were later swamped by successive waves of newcomers?

The discovery of Kennewick Man's great antiquity and European features set off a ferocious legal tussle. Scientists wanted to measure the bones and try to extract DNA from them. A group that practices an ancient European religion sued to gain access to the bones. But the strongest claim came from local Native American tribes. The 1990 Native American Graves Protection and Repatriation Act requires that the federal government contact any tribe having a "cultural affiliation" with bones found on federal property. If a tribe can produce any "geographical, kinship, biological, archaeological, anthropological, linguistic, folkloric, oral tradition, historical, or other relevant" evidence of their connection to the bones, the remains must be turned over to the tribe. A few days after the radiocarbon dates came back on Kennewick Man's finger bone, a coalition of tribes in Washington State filed a joint claim to have the remains given to them for reburial. As one Indian leader wrote at the time, "Our elders have taught us that once a body goes into the ground, it is meant to stay there until the end of time."

The federal agency with jurisdiction over the bones immediately agreed with the tribes. The bones had been found on land owned by the U.S. Army Corps of Engineers, which tries to maintain good relations with local tribes because it often has to work with them. In response to the Corps' decision, a group of eight scientists sued for the right to study the bones, arguing that they represented a unique national treasure. In September 2000, the Corps' parent agency, the Department of the Interior, sided with the tribes. Interior Secretary Bruce Babbitt wrote, "The geographic and oral tradition evidence establishes a reasonable link between these remains and the present-day Indian tribe claimants."

Questions of ancestry raise many difficult issues. Our DNA connects everyone to everyone else after a couple of thousand years, so determining who is an ancestor of whom is far from straightforward. If ancestry is associated with genes, a homeland, or a culture, who has the right to claim ownership of the past?

Genetics has already revealed important new information about the settlement of the New World and promises to deliver much more. But this research also has had an unexpected consequence. In other parts of the world, the DNA evidence generally has had a clarifying effect on our understanding of the past, cutting through at least some of the confusion generated by an inevitably incomplete archaeological record. That has happened in the Americas too, but the genetic evidence has also revealed the inherent complexities of the past. It has shown that history is much more complicated than we think.

In 1589 the Jesuit scholar José de Acosta, who lived and traveled widely in South America, proposed that Native Americans were descended from people who had migrated from Siberia. More than four hundred years later, Acosta's idea has held up pretty well. Perhaps 75 million people were living in North and South America when Columbus reached the New World in 1492. Most, perhaps all, of their ancestors have been shown to be people from Asia who made their way across what is today the Bering Strait. The questions — and the controversies — lie entirely in the details.

The single most contentious question concerns the dates of these migrations. At this point the only certainty is that people were living in the Americas by about 13,500 years ago, because the American archaeological record suddenly springs to life at that date. Sites throughout the two continents attest to widespread human occupation. Furthermore, many of the sites in North America, though not those in South America, have produced a particular kind of stone tool, a leaf-shaped arrowhead with a concave base known as a Clovis point, named for Clovis, New Mexico, where the first such artifact was found. Archaeologists have concluded that either the people at these widely separated sites interacted with one another so extensively that the technology of Clovis points spread quickly, or all of these groups were descended from ancestors who used and passed on the knowledge of how to make Clovis points.

For many years most archaeologists working in the Americas favored the second option. Clovis points appear in North America during a particularly suggestive period. Before about 14,000 years ago, walking south across what is today Canada was not possible because glaciers extended across North America from the Pacific to the Atlantic. No group of humans could have traveled over thousands of miles of heavily crevassed ice. Then, about 16,000 years ago, the glaciers began to recede. Within a thousand years or so a corridor had opened up just to the east of the Rocky Mountains, where the Canadian plains meet the foothills. This corridor initially must have been a forbidding passageway, with high walls of ice, endless marshes, and harsh winter winds. But eventually the passageway widened and dried, and by 14,000 years ago or so — experts still debate the exact date — it must have been passable.

A few hundred years later, Clovis points show up in North America. The opening of the corridor and the appearance of Clovis points have always seemed too close in time to be coincidental. Since the initial finds of Clovis points in the 1930s, archaeologists have supposed that people bearing Clovis technology came through the corridor and spread quickly across North America. Perhaps they were following herds of big game migrating south through the corridor. Never having seen humans before, the animals would have been an easy mark for hunters. In fact, many of the large prehistoric mammals of America went extinct about this time, though the warming climate, rather than successful hunters, may also have caused their demise.

What has always made this scenario particularly compelling has been the lack of evidence for the occupation of the Americas before the arrival of the Clovis people. Archaeologists have looked hard for such evidence, and over the years they have proposed dozens of pre-Clovis sites. But these claims have had a spectacularly bad track record. Almost without exception, the sites have turned out to have problems when examined closely. The "artifacts" found were actually produced by natural forces, or the site was much younger than supposed, or later artifacts had been mixed into earlier layers. As potential pre-Clovis sites were eliminated one by one, the conviction grew that the Clovis people must have been the continent's earliest inhabitants.

This "Clovis First" hypothesis eventually gained support from other scientific disciplines. In particular, researchers began to find evidence that discrete waves of people had moved into the Americas from Asia, and they tended to link the first of these waves with the Clovis people. One kind of evidence came from linguistics — and specifically from Stanford linguist Joseph Greenberg's proposal that all of the languages spoken by Native Americans at the time of European contact could be grouped into three large families. The most recent language family to enter the Americas, Greenberg said, is represented by about ten Eskimo-Aleut languages spoken by Arctic hunter-gatherers who spread across Alaska into northern Canada and Greenland within the last few thousand years. The next most recent consists of the forty or so Na-Dené languages spoken by people in northwestern North America and by their offshoots the Navajo and Apache, who migrated to the Southwest about a thousand years ago. The oldest grouping, encompassing almost a thousand current and extinct languages, is the Amerind family, according to Greenberg. Though the Amerind languages are extremely diverse, Greenberg thought he saw enough similarities to infer descent from a single proto-language. For example, in many Amerind languages, first-person pronouns begin with *n*, and second-person pronouns with *m*. (In Indo-European languages, these pronouns tend to begin with *m* and *t*, respectively.) Though Greenberg did not specify when these languages came to America or who was speaking them, a reasonable assumption was that Proto-Amerind was the language of the Clovis people.

A second form of evidence came from examinations of teeth. For the past four decades Christy Turner of Arizona State University has been leading an effort to trace the spread of modern humans across the globe using dental evidence. Turner and his students and colleagues examine tooth characteristics that only a dentist could love, with names such as Carabelli's cusp, double shoveling, and Y grooves. Based on an analysis of more than 9,000 fossil crania from North America, they concluded that the teeth of past Native Americans fall into roughly three categories. Skulls from areas where people speak Eskimo-Aleut languages form one cluster; fossils from what Turner has called the Greater Northwest Coast area (corresponding to Greenberg's Na-Dené language speakers) form another; and all other Native American fossil teeth comprise the third cluster. Moreover, in many respects the Na-Dené cluster lies between the Eskimo-Aleut and Amerind clusters, again lending support to the hypothesis of three waves of migration.

Finally, the available genetic evidence also pointed to successive waves of Asian immigrants. When these migrations were first studied, geneticists were not yet able to determine mitochondrial or Y-chromosome haplotypes. But the protein evidence seemed to lend support to a tripartite division of Native Americans, though the overlaps among groups were extensive.

In a seminal article in the journal *Current Anthropology* in 1986, Greenberg, Turner, and a University of Arizona geneticist, Stephen Zegura, laid out the case for what has become known as the three-wave model. The first people to enter the Americas, according to the authors, spoke a language ancestral to the Amerind family and spread rapidly through the New World. A wave of Na-Dené speakers entered several thousand years later, followed by a wave of Eskimo-Aleut speakers. "In our opinion, the most reasonable historical interpretation is that these three represent three migrations from Asia," the authors wrote.

Greenberg, Turner, and Zegura's 1986 paper has for many years been one of the most cited works in American archaeology. However, the bulk of the citations have come from researchers eager to poke holes in it.

The three-wave model did not depend on the Clovis people being the first to enter the Americas. Yet the model fits so well with the idea of big-game hunters marching through the glacier corridor that the two were inevitably conjoined. As a result, any cracks in the Clovis First hypothesis have inevitably cast doubts on the three-wave model.

Those cracks have widened dramatically in recent years. Several sites in North and South America now strongly suggest that people were in the New World well before the appearance of Clovis points. Near Charleston, South Carolina, small blades from a sandy hillside have been dated to a thousand years before Clovis. At the Meadowcraft rock shelter in western Pennsylvania, radiocarbon dates for the earliest occupation layer exceed 19,000 years. Near Richmond, Virginia, a scraper and small stone blades appear to come from an occupation layer more than 15,000 years old. Though these sites have their critics, each is beginning to look pretty solid.

But the strongest pre-Clovis evidence comes from a site known as Monteverde in south-central Chile. In an upland bog about thirty miles from the Pacific Ocean, a research team led by Thomas Dillehay from the University of Kentucky at Lexington has found stone and wood tools, fire pits, the remains of structures, and other signs of human habitation. Radiocarbon dating indicates that these artifacts are about 14,700 years old, more than 1,000 years older than Clovis. Though some aspects of Monteverde's dating remain controversial, the balance of opinion has shifted toward accepting it as a pre-Clovis site. And Dillehay says he has found evidence of even older occupations at nearby sites in Chile.

The case for pre-Clovis settlements is not yet airtight. That the archaeological record of pre-Clovis people remains so elusive is odd. Perhaps there weren't many of these people. Or maybe they used technologies based exclusively on wood and fibers that did not preserve well. In any case, the search goes on for a site that will prove, beyond any shadow of a doubt, that Clovis First must be relegated to the dustbin.

Meanwhile, the other lines of evidence offered to support the three-wave model likewise have come in for rough treatment. At this point, few linguists agree with Greenberg's claim that all of the native languages outside Alaska and northwestern North America belong to a single family. Though Greenberg's followers remain convinced, critics insist that these languages can be reduced to no fewer than a dozen or so subgroups. This controversy has been one of the nastiest in linguistics. As one reviewer of Greenberg's proposal put it, "The whole speculative venture should be abandoned. Indeed, the linguistic classification should be shouted down in order not to confuse nonspecialists."

The dental evidence has come under similar criticism. The correspondence between Turner's Greater Northwest Coast area and Greenberg's Na-Dené language area is admittedly rough. Furthermore, this area must have seen extensive mixing of new migrants from Asia with earlier occupants. Turner's critics have asked whether he would have seen the same dental patterns if he had not had the three-wave model in mind.

The genetic evidence has undergone the same critical scrutiny as have the linguistic and dental data. But in the case of genetics, new information has become available that has further highlighted some of the shortcomings of the three-wave model.

Geneticists have been using proteins to study the origins of Native Americans at least since the 1930s. But more detailed studies of DNA sequences began much more recently. In the early 1970s a young graduate student at Yale Medical School named Douglas Wallace became curious about the DNA in mitochondria. Today Wallace is director of the Center for Molecular Medicine at Emory University in Atlanta and one of the world's leading experts on mitochondrial genetics. But three decades ago very little was known about mitochondrial DNA. "It was an exciting time," Wallace told me recently when I visited him in his office overlooking Emory Medical School. "We worked out all the basic principles of mitochondrial genetics back then — maternal inheritance, the high mutation rate, sequence divergence."

In 1975 Wallace received his Ph.D. and moved to Stanford, where he continued his work on mitochondrial DNA as an assistant professor. At that time DNA sequences longer than a few hundred nucleotides could not be sequenced directly, so Wallace instead used some of the recently discovered tools of genetic engineering to examine variations in mitochondrial DNA. In particular, a new class of proteins known as restriction enzymes had been discovered that cut DNA at particular nucleotide sequences. For example, an enzyme called *Hpa* I cuts DNA every time it encounters the sequence GTTAAC. So if *Hpa* I is mixed with mitochondrial DNA from a particular individual, the enzyme will cut that DNA into a collection of segments of particular lengths that depend on the number of times the sequence GTTAAC appears. Now say that a different person has exactly the same mitochondrial

DNA sequence except for one nucleotide; one of the GTTAAC sites has mutated to GTCAAC. If that person's mitochondrial DNA is mixed with *Hpa* I, the enzyme will cut all of the GTTAAC sites but not the mutated site. As a result, the enzyme will produce one less DNA segment, and one of the segments in the digested mitochondrial DNA will be longer. In this way, using different restriction enzymes, Wallace slowly and methodically probed how mitochondrial DNA varies from person to person.

In the late 1970s Wallace began to compare mitochondrial DNA from people in different parts of the world. He wasn't interested primarily in history; his research has always focused largely on disease. A particular interest of his has been Leber's Hereditary Optic Neuropathy, or LHON, a disease that typically attacks the optic nerves of young adults and destroys their vision. Wallace was the first researcher to show that LHON results from mutations that impair the efficiency of mitochondria. As the mitochondria fail to produce enough energy to keep cells running, the cells start to die, and the hard-working neurons of the visual system are the first to go. By comparing the DNA of people from different parts of the world, Wallace hoped to figure out which mutations were responsible for the disease.

He found something that he did not expect. People from various parts of the world had distinct patterns of mutations in their DNA. Most Africans had one set of mutations, most Asians another, and most Native Americans a third. "It never would have occurred to me that variation would be distributed that way," he says. Wallace decided that the mitochondrial DNA patterns had to result from human history, and he set out to discover why.

All of this work was done without the knowledge that Allan Wilson's group at Berkeley had started their research on mitochondrial DNA and human origins. "I didn't learn that Allan Wilson's student Becky Cann was working on mitochondrial DNA until I visited Berkeley to give a seminar," Wallace recalls. He and his colleagues at Stanford were in the midst of a worldwide study of mitochondrial haplogroups when Wilson's team made headlines with mitochondrial Eve. "They were very good at encapsulating in a single idea what we all were doing."

After Wilson's death in 1991, Wallace continued his exploration of mitochondrial haplogroups, and many of the conventions used today date from these early investigations. For example, Wallace and his stu-

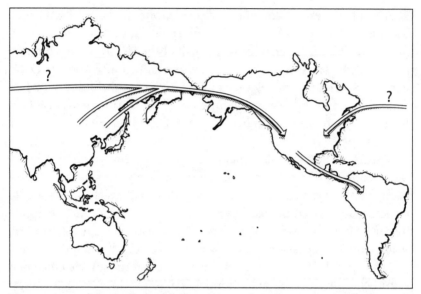

The ancestors of the earliest Americans migrated to the New World from central and eastern Asia, according to comparisons of Asian and Native American mitochondrial DNA and Y chromosomes. One type of mitochondrial DNA found in some Native Americans, haplotype X, seems to have originated in Europe more than 10,000 years ago, traveling across either northern Asia or the North Atlantic to reach the Americas.

dents were engaged in a study of Native American mitochondria when they decided that the terms used to refer to mitochondrial mutations were too cumbersome. They decided to call the four major haplogroups found in Native Americans A, B, C, and D. As they worked their way from North America back to Asia, Europe, and Africa, they began giving the other major haplogroups letter names — L for the major group in Africa, M for that in Asia.

Wallace's lab also initiated a much more detailed study of individual haplogroups. For example, their study of Native Americans clearly showed that the American versions of haplogroups A through D were derived from a related set of A through D haplogroups in Asia. Wallace and his colleagues then looked in Asia for the people with mitochondrial DNA most closely related to that of Native Americans today. Surprisingly, these closest mitochondrial relations did not live right across the Bering Strait on the Chukchi Peninsula. Rather, the people most closely related to Native Americans through haplogroups A, C, and D lived near the border of Mongolia and Siberian Russia, around the Altai Mountains and Lake Baikal. The closest relations for haplo-group B lived even farther away, along the coast of China and South-east Asia. (More recent work on the Y chromosome has revealed simi-larly diverse origins for North American males, ranging from central to eastern Asia.)

This comparison of American and Asian haplotypes was important for another reason. The differences within haplogroups can indicate how long the groups have been separated, but here a problem soon arose. The American haplogroups were too distinct from their Asian relatives to have arrived in the Americas as recently as 13,000 years ago. In fact, their diversity seemed to indicate that haplogroups A, C, and D had been in the Americas for more than 20,000 years. Haplo-group B appeared somewhat younger, having arrived in the Americas perhaps 15,000 years ago. Still, these dates seemed to say that humans had been in the Americas long before the arrival of the Clovis people.

An even greater surprise was on the way. Wallace's lab began to find an unusual Native American haplogroup — known as haplogroup X — particularly among Algonquian-speaking groups living around the Great Lakes. It was not that these mitochondrial sequences were mys-terious — X was simply the next letter in the alphabet when Wallace's group was naming haplogroups. The surprise was that haplogroup X had been found previously only in Europe, in Druze, Italian, and Finnish populations. Despite extensive searches, haplogroup X had never been found among Asians.

At first Wallace thought that the Xs came from Europeans who had mixed with Canadian tribes during the 1600s and 1700s. But the Na-tive American Xs differed too much from the European Xs to be sepa-rated by just a few hundred years. Indeed, in a study of Native Ameri-can skeletons from a fourteenth-century graveyard in Illinois, two of the skeletons carried unmistakable traces of haplogroup X. Given the amount of divergence between the European and American Xs, the haplogroup must have been in the Americas for more than 10,000 years. Wherever X came from, it couldn't have come from a modern European.

The settlement of the Americas has always attracted bizarre archaeo-logical hypotheses. One proposes that both the Egyptians and the Mesoamericans are descended from the ancient inhabitants of Atlan-tis, because they both built pyramids aligned with the stars. The *Book of Mormon* says that ancient Israelites migrated to the New World and helped build the great cities of Central America. In his bestseller *Chariots of the Gods,* the Swiss writer Erich Von Däniken wrote that extraterrestrial visitors shaped the Aztec and Inca civilizations.

Genetic evidence cannot rule out any of these ideas, no matter how absurd. But geneticists can say this: if any Africans, Australians, Europeans, or extraterrestrials made it to the Americas before Columbus, they didn't have much luck breeding with the locals. No mitochondrial DNA from haplogroups of people living anyplace other than Asia have been found in the Americas, with one possible exception: haplogroup X might represent an early arrival of people from Europe. But even that haplogroup could have come from Asia. People from Europe and Asia have mingled across the northern tier of the continent for thousands of years. During that time a haplogroup found mostly in Europeans today could have existed in an Asian population. This population could have carried the haplogroup into the Americas along with haplogroups A through D. In fact, the recent discovery of a few people with haplotype X living near Lake Baikal in southern Siberia suggests that this is probably what happened.

But another intriguing possibility remains. For many years archaeologists have wondered if, during the last Ice Age, Stone Age Europeans could have made their way along the edge of the ice sheet bordering the northern Atlantic Ocean from Europe to the Americas. Such a voyage would have been incredibly difficult. Most of the route would have paralleled immense, calving glacial walls. The water would have been filled with sharp, jostling chunks of ice. The route would have offered very few places to land and get food or water. Given that people abandoned even northern Europe during the Ice Age, the idea that they floated alongside an ice sheet for thousands of miles seems extremely farfetched.

Still, such a migration would solve an archaeological mystery. For many years archaeologists have pointed out that Clovis points look very similar to some stone tools known as Solutrean points made in Spain and Portugal between 22,000 and 16,500 years ago. A 3,000-year gap separates the last known Solutrean points and the earliest Clovis points, with no obvious explanation for the hiatus. But if Solutrean points did make it to the Americas, they could have been the inspiration for Clovis points. This might also explain why Clovis points have never been found in Asia. And Clovis points seem to show up first in southern North America and then spread north, not the other way around. "I'm very willing to consider an Atlantic crossing," says Wallace.

Geneticists may be able to help resolve this debate over the next few years. As they continue their study of people with haplogroup X, they gradually will uncover genetic differences among them — more detailed mitochondrial mutations that separate all the Xs into subgroups. If the X haplotypes in eastern Europe and Asia are more like those in the Americas, these mitochondria probably came across Asia. But if the X haplotypes of western Europe are more similar to the American variety, then the first people to carry haplogroup X to the Americas may have come across the northern Atlantic.

New genetic evidence also could produce better dates for the arrival of various groups in the Americas. At this point the Clovis First hypothesis is in serious trouble. But if people were living in the Americas before the opening of the ice corridor, the question is how they arrived. Before the peak of the last Ice Age, a pathway from Asia between the glaciers may have been open between about 35,000 and 30,000 years ago. These dates had always seemed unreasonably early for the arrival of the first Americans. But they correspond fairly well with the genetic data. Under this scenario, the first wave of colonists from Asia would have entered the Americas well before 20,000 years ago. Then the ice corridor would have shut, isolating this first group from the ancestral populations in Asia. After the ice corridor opened again, new migrations would have occurred, including those of the Na-Dené and Eskimo-Aleut speakers.

Then again, maybe Stone Age Asians reached the Americas by boat along the coastline. This trip would not have been much easier than a journey along the northern Atlantic ice sheet, and it would have taken even longer. However, a circum-Pacific route would have had more nonglaciated landing spots along the way. And early modern humans were obviously skillful at traversing coastlines, as is evident from the early occupation of Australia. Proving that the first Americans traveled by boat would be hard. Most of their shoreline campsites would be under water now, and any boats they used would long since have decayed into dust. But the route remains plausible.

An early arrival of Asian immigrants would solve another archaeological mystery — that of the skulls. The European features of skulls such as Kennewick Man are hard to explain if their ancestors came from Asia at the end of the last Ice Age, when Asians had a more characteristically Mongoloid appearance. But if the first migrants came earlier, they would have looked more European because the Mongoloid traits had not yet developed in Asia. After the traits typical of today's eastern Asians appeared, later migrations would have brought those facial features to the New World. Over time, the earlier and later populations would have mixed, producing the varied appearances of the groups that occupied the Americas when Europeans arrived in the fifteenth century.

Maybe it doesn't really matter exactly who settled the Americas. Yet people remain fascinated by the subject, which I think reflects the different ways in which we view history. People have many connections to the past, and in the case of the Americas those connections are especially complicated. Some Native Americans simply trace their ancestry to the occupants of the New World before Columbus, although the majority also have European-American or African-American ancestors. Other people may be tempted to justify the appropriation of Native American lands by claiming that they were colonists, too. Some look to the ancient Americans for a spiritual link to a simpler world. Others see in precontact Native Americans an inferior culture doomed to fail.

As is always the case with genetics and history, we interpret the past selectively, picking out those features that accord with our worldviews. Kennewick Man is a perfect example. So long as he had children before he met his demise on the banks of the ancient Columbia River, he is probably a direct ancestor of everyone alive today. Some of his descendants may have moved back over the Bering Strait into Asia, where they would have become the ancestors of all Asians, and eventually of all Asians, Europeans, and Africans. In that case, when Columbus met the Arawak Indians on the shores of San Salvador, he was encountering his own distant cousins. Everyone on that beach was descended from Kennewick Man.

People create their own pasts by acknowledging what they choose to acknowledge. In the 1960 U.S. census — the first that allowed people to classify themselves by racial category — just over 500,000 people identified themselves as Native Americans. By the 1980 census more than 1.4 million said that they were Native Americans. And in the 2000 census, which for the first time allowed people to identify themselves as belonging to more than one race, more than 4 million Americans marked "Native American" on their census form.

Clearly the Native American population has not grown eightfold over the past two generations. Rather, more people have chosen to recognize or claim some degree of Native American ancestry. In part those choices reflect financial considerations involving casino revenues, affirmative action programs, and federal land payments. But they also reflect different perceptions of the past. A hundred years ago, most Americans who had the option of overlooking a Native American ancestor probably did so. Today, as the stigma of such ancestry has receded, many people are embracing their past.

Greater recognition of our genetic connectedness will not make the social and political issues we face less vexing. Societies must still decide whether and how to compensate people whose ancestors were not accorded the rights now extended to all. More broadly, individuals and governments alike must choose whether to ignore or confront a history of prejudice and injustice directed against specific groups. These are difficult questions that do not have simple answers. But they cannot be addressed honestly without acknowledging the realities — and the complexities — of our biological histories.

12

The Burden of Knowledge

Native Americans and the Human Genome Diversity Project

> The end of man is knowledge, but there is one thing he
> can't know. He can't know whether knowledge will save him
> or kill him. He will be killed, all right, but he can't know
> whether he is killed because of the knowledge which he has
> got or because of the knowledge which he hasn't got and
> which if he had it, would save him.
>
> — Robert Penn Warren, *All the King's Men*

BY THE EARLY 1990s it was becoming clear that genetics research had
the potential to reveal aspects of our history that had seemed lost forever. The discovery that mitochondrial Eve lived fewer than 200,000
years ago demonstrated the recency of our common ancestry. The
identification of mitochondrial haplogroups in various parts of the
world was beginning to unearth ancient migration patterns. And the
newly developed tools of genetic engineering were enabling researchers to study variation in ways never possible before.

At Stanford University, Luca Cavalli-Sforza realized that the pieces
were falling into place to achieve a lifelong dream. With four colleagues, including Allan Wilson at Berkeley, he wrote a paper entitled
"Call for a Worldwide Survey of Human Genetic Diversity." Published
in the journal *Genomics* in 1991, the article had an urgent tone. "The
genetic diversity of people now living harbors the clues to the evolution of our species, but the gate to preserve these clues is closing rapidly," Cavalli-Sforza and his coauthors wrote. The best way to understand the history of modern humans, they said, is to study groups that

have been relatively isolated for thousands of years, such as the Bushmen or Australian aborigines. But many of these groups are disappearing because of famine, disease, and war. Others are merging with their neighbors, scrambling the genetic signals needed to reconstruct their history. "It would be tragically ironic," the article concluded, "if, during the same decade that biological tools for understanding our species were created, major opportunities for applying them were squandered."

The imminent loss of so much of the world's human biodiversity demanded an immediate response, Cavalli-Sforza and his colleagues said. Blood samples should be gathered from individuals living in populations of special interest throughout the world. White blood cells in the samples then could be converted into laboratory cell lines, which would provide a permanent supply of DNA for anthropological and biomedical projects. Such a project — which soon became known as the Human Genome Diversity Project or HGDP — would produce what Cavalli-Sforza termed "enormous leaps in our grasp of human origins, evolution, prehistory, and potential."

Cavalli-Sforza did not expect that his proposal would generate great objections. In more than thirty years of studying genetic variation around the world, he had always found that people were willing to participate in his projects and were interested in the results. The HGDP would simply take that research to a new level. Looking at many more people and many more genetic markers would enable the genetic study of human history to achieve its full potential.

This time Cavalli-Sforza was mistaken. The proposal did not meet with an enthusiastic response from people eager to learn about their history. On the contrary, it loosed a flood of controversy. Aboriginal groups in the United States and elsewhere said the HGDP would steal their genes, destroy their culture, and even contribute to genocide. Academic critics claimed that the project would encourage racist thinking by oversimplifying issues of great complexity. The idea that genetically isolated groups existed and were somehow more pristine than others, and that they needed to be studied quickly before they were destroyed, came in for particularly harsh treatment. "The idea of studying human genetic diversity is a good one, but the way that Cavalli-Sforza has conceptualized it has problems at all levels," said Jonathan Marks, an anthropologist at the University of North Carolina at Charlotte.

The reactions baffled Cavalli-Sforza, who had always considered himself a leader in the struggle against racism. In the 1960s he worked closely with groups in Africa and wrote fondly of his friendships there. In the 1970s he participated in public debates with the Stanford physicist William Shockley, disputing Shockley's racist ideas. In his magnum opus *The History and Geography of Human Genes,* written with Paolo Menozzi and Alberto Piazza, he wrote an entire section on the "scientific failure of the concept of human races." From the beginning of his career, he thought that the study of human genetic variation would help to end racism, not inflame it.

In proposing the HGDP, Cavalli-Sforza placed himself at the center of the dilemma that has haunted the study of human genetic differences. The only way to understand how similar we are is to learn how we differ. Yet studies of human differences can seem to play into the hands of those who would accentuate those differences. Researchers might claim that the genetic differences they identify among groups have no biological significance. Yet simply by dividing humans into categories — whether sub-Saharan Africans, Jews, Germans, or Australian aborigines — geneticists can inadvertently reinforce the distinctions they would seek to minimize. How to resolve this dilemma is one of the most difficult problems facing the study of human genetics.

Cavalli-Sforza was working from an existing model when he proposed the Human Genome Diversity Project in 1991. The year before, a consortium of governments and private organizations had launched the Human Genome Project. This project, too, had initially generated controversy, although of a quite different type. When it was first proposed in the mid-1980s, many biologists viewed it as a colossal waste of time and money. The goal of the project — determining the sequence of all 3 billion nucleotides in a randomly chosen set of human chromosomes — would mean sequencing the 98 percent of the genome that has no obvious function as well as the 2 percent that does. Furthermore, the utility of the raw sequence was far from obvious. An endless list of nucleotides would not tell biologists how genes exerted their effects. To some critics, sequencing the genome was like "viewing a painting through a microscope" — you'd see the brush marks but completely miss what the painting was trying to show.

Gradually most skeptics were won over. A report from the National Research Council of the National Academy of Sciences recommended that the project go forward. The heads of both the National Institutes of Health and the Department of Energy (the latter has supported genetics research since it began studying the survivors of the Hiroshima and Nagasaki explosions) lobbied forcefully for the program — so long as new funds were made available to run it. When the project finally got under way in 1990, the organizers thought it would cost about $3 billion and take fifteen years to complete. Spurred by competition from the private sector, the project was completed faster than had been expected. By the spring of 2001 most of the sequence had been determined and, with great fanfare, was released to the public.

From Cavalli-Sforza's perspective, the Human Genome Project was a good idea, but it didn't go far enough. The sequenced DNA was a pastiche of chromosomal fragments from various unidentified donors, and the project produced just a single generic sequence to represent all of humanity. In that respect, the project is more accurately *a* Human Genome Project rather than *the* Human Genome Project. To trace human history, Cavalli-Sforza needed to know how DNA sequences vary among people from different parts of the world, and he needed to match those sequences with the people who provided them.

The HGDP was designed to fill those gaps. As plans for the project took shape, the goal became to collect DNA from several hundred distinct groups, including many indigenous groups. Immortalized in cell lines, this DNA would be used primarily to study human history, but it would also be a resource for medical researchers investigating the connections between genetics and disease. And according to the project's planning document, by demonstrating the nature of the genetic differences and similarities among people, the HGDP "would help to combat the widespread popular fear and ignorance of human genetics and will make a significant contribution to the elimination of racism." No one, Cavalli-Sforza and his colleagues reasoned, could be opposed to that.

In 1993 an odd-looking document appeared on the desks of the HGDP's organizers. Under the heading "RAFI Communiqué" was the title "Patents, Indigenous Peoples, and Human Genetic Diversity." An artful combination of analysis and innuendo, the document was unambiguous in its conclusion: "The Human Genome Diversity Project should immediately halt any collection efforts."

The campaign against the HGDP marked the first foray into human genetics for RAFI — the Rural Advancement Foundation International. A small organization based in Canada, RAFI had previously targeted corporations that removed indigenous plants from developing countries, repackaged their genetic material in hybrid seeds, and then offered the patented seeds to Third World farmers for exorbitant prices. Now it accused the HGDP of a similar form of "biopiracy." The DNA of indigenous people would be mined for valuable information, which pharmaceutical companies would then use to make drugs far too expensive for Third World people to buy.

The actions of the U.S. government in the early 1990s seemed to bear out RAFI's concerns. Federal agencies had applied for patents on cell lines containing DNA from several Third World groups. The patent applications, all aimed at medical uses, were unrelated to the HGDP, which repeatedly dissociated itself from any commercial ventures. But opponents of the project eagerly conflated the uproar over the patents with what they called the "vampire project."

Native American groups were particularly critical of the HGDP. To them the project appeared to present only risks, not benefits. They worried about the uses of the medical information generated by genetics research. Even more, they worried about the anthropological results of such research. Soon Native Americans and other groups representing indigenous peoples put together a formidable list of objections to the HGDP.

The first objection actually applies to all genetics research. Why should a group participate if the research is not likely to generate any obvious benefits for the group? Genetic tests rarely deliver good news; much more often, they reveal that a person has inherited a genetic defect that raises the risk of suffering from a disease such as breast cancer or Alzheimer's. If that result leads the person to make lifestyle changes that reduce the likelihood of getting the disease, then a genetic test may be worthwhile. But research often focuses on diseases for which no immediate cure or preventive measure is available. In that case, people have to ask themselves why they should submit to a genetic test that could place a black cloud over their head for the rest of their life. For example, the majority of people at risk of inheriting the mutated gene responsible for Huntington's disease — which causes nerve degeneration and eventual death after the age of about forty — decline to be tested for the mutation. They would rather live with uncertainty about the disease than with the knowledge that the disease is inescapable — even though the test might show that they do not have the mutation.

The second objection to studies of genetic variation concerns public perceptions of groups. Genetics research is showing that some groups have an unusually high frequency of genetic variants associated with a disease. These variants rarely have a major effect on the health of the group as a whole, given the complexity of the links between genes and disease. Nonetheless, groups have worried that such a finding could stigmatize all of its members, regardless of the size of the effect or the percentage of the group having the variant. For example, say an Indian tribe had a higher than average percentage of a genetic variant that increases the risk of diabetes by 5 percent. Most new cases of diabetes in the tribe would still be caused by environmental factors or by the actions of other genes, given the relatively minor influence of the gene being studied. But if the entire tribe were labeled as particularly prone to diabetes, its members could have trouble finding insurance coverage or jobs.

A third concern involves a group's conception of itself. "The creation stories of Native Americans do not involve migrations across the Bering land bridge," says Debra Harry, a member of the Northern Paiute Nation and the director of an organization called the Indigenous Peoples Council on Biocolonialism. "Our creation stories relate us directly to our aboriginal territories." The significance of these beliefs extends well beyond the purely cultural. In the past the U.S. government has sometimes denied Native Americans tribal status because they were unable to establish clear lines of descent from an ancestral group. "Native Americans are deeply and rightly suspicious about government-sanctioned scientists purporting to tell them about their origins," Harry points out. "The colonial agenda has been, and continues to be, to dispossess indigenous peoples of their resources. Our basic humanity and human rights are ignored in favor of genocidal practices and policies that 'get rid of the Indian problem.'"

Fourth, in many native cultures, the idea of taking DNA and storing it in a freezer is sacrilegious. "DNA is not ours to manipulate, alter, own, or sell," says Harry. "It was passed on from our ancestors and should be passed on to our children and future generations with its full integrity." Native Americans and other groups tapped into a well-spring of public concern when they pointed out that the creation of cell lines would mean that the full complement of a person's DNA would be stored in a repository indefinitely. The distance is short from that image to the scare stories that have circulated widely on the Internet — for example, that government-sponsored scientists want to use people's DNA to create a race of zombies.

Finally, financial issues have been a stumbling block. If information about genetic variation has so little commercial value — as the organizers of the HGDP have repeatedly claimed — why have so many cell lines and genes been patented?

The organizers of the HGDP responded to all the objections. They insisted that no individual or group could be harmed by the information they were gathering. They pointed out that studies of human history were unlikely ever to make money. But the organizations representing indigenous peoples often seemed — at least to the scientists — more interested in confrontation than in discourse. "These groups have to support themselves somehow," Cavalli-Sforza has said, "and they've decided to use us." At the same time, the critics were more adept than the geneticists at questioning the judgment of their opponents. "Nice, smart, liberal — but naive and a bit arrogant — white guys firmly ensconced at reputable universities," is how a RAFI organizer once described Cavalli-Sforza and his colleagues to me.

Realizing that the HGDP was slipping from his grasp, Cavalli-Sforza turned to a Stanford law professor named Hank Greely to help represent the project. Over the next few years Greely, who had written widely about the ethical dimensions of genetics, pushed the geneticists to examine issues of consent and equity that had never been dealt with before. But by the fall of 1993, the controversy was spiraling out of control. Dozens of organizations representing indigenous peoples were attacking the HGDP. They decried the spending of millions of dollars to collect blood from groups that were disappearing because of poverty, disease, and official neglect. Project organizers responded by suggesting that they could provide medical assistance to participating groups. Then a new objection arose: promises of health care were being used to bribe indigenous peoples for their DNA.

For Greely the low point came in December 1993, when he traveled to Quetzaltenango, Guatemala, to represent the project at a meeting of the World Council of Indigenous Peoples. Scheduled to speak twice at the meeting, Greely spent most of his second session listening to angry speeches from the audience. By the end of the conference he was standing at the podium being accused of being a CIA agent. "I had the opportunity to respond after each comment and almost always did so," he later recounted in a report about the meeting.

> I noted that this was not a project about indigenous peoples, but about all the world's peoples. . . . I stressed that populations would not be involved in the project unless they wanted to be. I agreed that the West and western science had done terrible things to indigenous peoples, but said that science could also do some good things. I urged that our project was different — that we were trying to do things right.

His protestations were to no avail. "My statements were uniformly either ignored or dismissed as lies," he recalls. "The only faintly positive feedback from the crowd was that two speakers complimented me for my courage."

Many of the issues raised by Native Americans and other indigenous peoples are not new. They have come up repeatedly as researchers have learned more about the connections between genetics and disease. In the 1960s, for example, research into sickle cell anemia led to a comparable controversy. The disease is caused by mutations in the genes that direct the production of hemoglobin. People who inherit the mutations from both their mother and father typically suffer from extremely painful and debilitating episodes during which their cells do not receive enough oxygen. However, if a person gets the mutation from just one parent, he or she generally does not suffer from the disease but rather is a carrier. If he or she marries another carrier, each of their children has a one-in-four chance of inheriting mutations from both parents and suffering from the disease.

In the 1960s, geneticists developed tests that made it possible to identify carriers of sickle cell mutations. Several states responded by passing laws requiring African Americans to be tested. But African Americans are not the only people who suffer from the disease. Other people with ancestors who lived in tropical parts of the world also have elevated frequencies of sickle cell mutations. Moreover, the results of the tests were widely misinterpreted. Some African-American carriers lost their jobs or insurance coverage, even though they had no chance of contracting the disease. Though the laws were later repealed, hard feelings about the discriminatory effects of the testing have lingered among many African Americans.

The traditional way of dealing with such issues has been through informed consent. People who are considering a genetic test need to know what they are getting into. They need to know about the potential benefits and risks of a test. They need to know its accuracy and what will be done with the results. They need to know how the results could affect family members, employers, insurers, and others. Only after people are fully informed about a test and its possible consequences are they asked whether they want to be tested.

If done conscientiously, obtaining informed consent can be a time-consuming and expensive process. People who know both genetics and ethics must spend considerable time with potential subjects explaining, educating, and listening. And the job is not done once blood is drawn. The results of the test and the consequences need to be explained.

Informed consent traditionally has focused on individuals. But studies of genetics and history are typically focused on groups rather than individuals. The individuals tested are usually anonymous in such a study. The identifier on the sample would simply read "Berber" or "Druze" or "Apache."

Hank Greely's major contribution to the Human Genome Diversity Project has been to broaden consent to take this wider focus into account. During the mid-1990s he and a group of other ethicists and geneticists put together a document called the "Model Ethical Protocol," which was designed to serve as a guide for the collection of samples under the HGDP. Besides reviewing the various responsibilities incumbent on geneticists in studying human genetic variation, the protocol insists on what it called group consent. "The HGDP requires that researchers . . . show that they have obtained the informed consent of the population, through its culturally appropriate authorities where such authorities exist," it states. Who speaks for a group is not always obvious, Greely admits. But when a representative body can be identified, it should be consulted.

In theory, group consent sounds like a great idea. If a group is the subject of research, then logically its members should have some say over whether their genes are studied. But in practice, group consent is not so simple.

The most immediate problem is the one identified by Greely — deciding who represents a group. Say a researcher wants to conduct a study of Americans with Italian ancestors. Various cultural organizations represent Italian Americans, such as the Italian Cultural Society. But if geneticists walked into the society's headquarters in Washington, D.C., with a proposal to study the genes of Italian Americans, the society's employees would probably laugh in their faces. Such organizations have no real interest in studying the genetic history of people with Italian ancestors. They serve the interests of people who claim, accurately or not, to have Italian ancestors.

Geneticists could approach the Italian government for permission to conduct their study. But the Italian government has no authority over the actions of people living in the United States. They could talk to officials in the Catholic Church or to the descendants of long-established Italian families. But in the end no one organization represents the interests of Italian Americans.

However, other groups, including many Native American groups, do have recognized governmental structures. Under the Model Ethical Protocol, researchers wishing to study a particular American tribe should obtain the consent of the tribal government. But other groups within Native American communities also have great authority, including community elders, religious leaders, and leading families. Unless these groups are involved, permission from a tribal government may have little legitimacy.

Other issues are even thornier than deciding who speaks for a group. Groups are defined through social, political, and cultural processes; they are not defined genetically. But by seeking approval from a group, geneticists seem to imply that a group exists as a genetic entity first and as a cultural entity second. Geneticists have other options. They could organize their results geographically, for example. They could say that a particular set of samples was from individuals living near 35.5 degrees north latitude and 110 degrees west longitude. But then anyone could simply go to a map and figure out which group the researchers were talking about. (The coordinates I just gave are for the homelands of the Hopi Indians in Arizona.)

The problem with using socially defined groups to present genetic results is that it implies that these groups somehow are much more biologically distinct than they are. All of the complicated relationships between culture and genetics are ignored. People read into the results the conclusions generated by their prejudices.

Perhaps the problem lies not in the implementation of group consent but in the nature of consent. After all, the whole process of obtaining consent, whether from individuals or groups, can seem strikingly one-sided. A team of scientifically trained professionals comes into a consultation room or meeting hall. There they tell prospective research subjects what they, the scientists, think the subjects should know. Time might be allotted for a few questions. But then the subjects have to make a thumbs-up or thumbs-down decision — whether to participate or not. Plenty of evidence demonstrates that this process doesn't always work as intended. Individuals rarely receive all of the information they need. Many are eager for advice, whether from other prospective subjects or from the people administering the test. Geneticists often present the information in such a way that only one decision is logically possible, and many people are hesitant to ask questions in such settings. Some ethicists ask whether informed consent is an unattainable ideal.

Someone who has thought deeply about these issues is Morris Foster, an anthropologist at the University of Oklahoma at Norman. Foster grew up in Oklahoma before attending graduate school at Yale. Since returning to Norman he's been involved in several large genetic studies in the state's Native American communities. In the process he has arrived at a new understanding of how best to conduct genetic research on groups.

Foster is an advocate of what he calls community review. The basic idea is that a group should be seen not as the subject of research but as a partner in it. In some cases this might involve no more than having informal discussions with the members of a community before a project starts. In other cases it might amount to negotiating a formal agreement with a study community — essentially meeting the demands of group consent. The critical aspect of community review is that information flows both ways so that the interests and concerns of the group shape the design of the study. "Community review requires establishing a dialogue," Foster says. "The process doesn't really work unless you can get the two groups talking to each other."

Several years ago, Foster helped organize a study of diabetes in the Oklahoma Apaches. The main decision-making body for the tribe is the five-member Apache Business Committee (ABC), which is elected by tribal members. As a first step, the ABC held a series of public meetings in which researchers explained their goals. The committee then established a review board of community members to serve as an intermediary between researchers and members of the tribe. Involving families and other private groups in decisions was critical, says Foster, because many everyday decisions among the Apaches in Oklahoma are made at the family level. The research had to be discussed by individual families before the researchers could be sure that the community was fully engaged.

In the end the tribe strongly favored moving forward with the research, largely because diabetes is such a debilitating problem for so many of its members. But the community and the researchers also established a set of ground rules. The community review board has sixty days to review manuscripts produced from the study. If the board objects to the use of the tribe's name in the manuscript, it can request anonymity in any published paper. In the unlikely event that the project generates commercially important information, a portion of any funds generated will be returned to the tribe for health and education programs.

Such provisions inevitably impose constraints on researchers. But these kinds of arrangements are the only way to move forward with large-scale genetic studies, Foster believes. Community review allows individuals to identify possible risks, discuss issues among themselves, and arrive at decisions that reflect a collective judgment. The review process builds a relationship of trust that is a key element of success. "Many more communities are becoming aware of these issues as they're approached by researchers and asked to participate in genetic studies," Foster says. "At some point, researchers are going to find it difficult to enroll communities unless they engage in this process."

The Apaches placed one other constraint on the study of their genes: researchers cannot use the results to explore the history of the tribe. As a group, the Apaches of Oklahoma do not wish to establish their genetic relationship to past peoples or to other groups of Native Americans.

The ethical requirement that research benefit the individuals involved in it can be a challenge in studies of genetics and history. Identifying the advantage to a group in knowing more about its own past is not always easy. In the abstract, all people will benefit from knowing more about the history of our species, especially given the message of human unity emerging from that history. But these generalized benefits may not be enough to outweigh the risks a group sees in a proposed research project, especially if that group has routinely been oppressed by the broader society.

At the same time, groups are usually interested in research that produces health benefits, especially if it targets diseases that are prevalent in the group. The issue then becomes whether the study of a group's genetic predispositions to disease can be strictly separated from the study of that group's genetic history. Many geneticists think not. They point out that the genetic variants involved in a disease spread through a population as a result of historical events. Therefore, the only way to understand the distribution of those variants within a group is to study its history.

This proposition is now being put to the test. Government agencies and private companies are compiling huge databases describing the genetic variants of anonymous donors from specific human populations. They intend to use these data primarily to study genetic contributions to disease. But the same data can be used to reconstruct, in ever greater detail, the genetic history of human groups.

This is a story that does not yet have a tidy ending. Because of the controversy over the Human Genome Diversity Project, the U.S. government has so far granted the project relatively little funding. Yet research into the genetic differences among individuals and groups is thriving. Powerful forces, both commercial and personal, are driving the collection of vast amounts of genetic information. If genetic databases rigidly separate DNA variation data from the social identities of individuals, the genetic study of human history will continue to advance at its current pace. But if ways can be found to link the biomedical and anthropological applications of genetics, the next ten years will see unprecedented advances in our understanding of human history.

VI

The World

13

The End of Race

Hawaii and the Mixing of Peoples

> He loved everything, he was full of joyous love toward
> everything that he saw. And it seemed to him that was
> just why he was previously so ill — because he could
> love nothing and nobody.
>
> — Hermann Hesse, *Siddhartha*

ON THE MORNING of November 26, 1778, the 100-foot-long, three-masted ship *Resolution,* captained by the fifty-year-old Englishman James Cook, sailed into view off the northeast coast of the Hawaiian island of Maui. The island's Polynesian inhabitants had never seen a European sailing ship before. The sight of the *Resolution* just beyond the fierce windward surf must have looked as strange to them as a spaceship from another planet. Yet they responded without hesitation. They boarded canoes and paddled to the ship. From atop the rolling swells they offered the sailors food, water, and, in the case of the women, themselves.

One can easily imagine the contrast: the European sailors — gaunt, dirty, many bearing the unmistakable signs of venereal disease — and the Polynesians, a people who abided by strict codes of personal hygiene, who washed every day and plucked the hair from their faces and underarms, whose women had bodies "moulded into the utmost perfection," in the words of one early admirer. At first Cook forbade his men to bring the women on board the ship "to prevent as much as possible the communicating [of the] fatal disease [gonorrhea] to a set of innocent people." In the weeks and months to come, as the *Resolu-*

tion lingered offshore, Cook was far less resolute. Toward the end of 1779, the first of what are today called *hapa haoles* — half European, half non-European — were born on the island of Maui.

The nineteenth-century stereotype of the South Pacific as a sexual paradise owes as much to the feverish imaginations of repressed Europeans as to the actions of the Polynesians. The young women who swam out to the ships in Hawaii, Tahiti, and other South Pacific islands were from the lower classes, not from the royalty, which carefully guarded its legitimacy. Many were training to be dancers in religious festivals. They would rise in status by exchanging their sexual favors for a tool, a piece of cloth, or an iron nail.

The Polynesians paid dearly for their openness. At least 300,000 people, and possibly as many as 800,000, lived on the Hawaiian Islands when Captain Cook first sighted them (today the total population of the state is about 1.2 million). Over the course of the next century, diseases introduced by Europeans reduced the native population to fewer than 50,000. By the time the painter Paul Gauguin journeyed to the Pacific in 1891, the innocence that Europeans had perceived among the Polynesians was gone. "The natives, having nothing, nothing at all to do, think of one thing only, drinking," he wrote. "Day by day the race vanishes, decimated by the European diseases. . . . There is so much prostitution that it does not exist. . . . One only knows a thing by its contrary, and its contrary does not exist." The women in Gauguin's paintings are beautiful yet defeated, without hope, lost in a vision of the past.

Today visitors to Maui land on a runway just downwind from the shore where Captain Cook battled the surf eleven generations ago. Once out of the airport, they encounter what is probably the most genetically mixed population in the world. To the genes of Captain Cook's sailors and the native Polynesians has been added the DNA of European missionaries, Mexican cowboys, African-American soldiers, and plantation workers from throughout Asia and Europe. This intense mixing of DNA has produced a population of strikingly beautiful people. Miss Universe of 1997 and Miss America of 2001 were both from Hawaii. The former, Brook Mahealani Lee, is a classic Hawaiian blend. Her ancestors are Korean and Hawaiian, Chinese and European.

Bernie Adair — who was selling candles at a swap meet in Kahului, Maui's largest town, when I met her — told me that her family's history was typical. Adair, whose ancestors came to Hawaii from the Philippines, married a Portuguese man in the 1960s. In the 1980s their daughter Marlene married a man of mixed Hawaiian, Chinese, and Portuguese descent. Adair's granddaughter Carly, peeking shyly at me from under a folding table, therefore embodies four different ethnicities. "These children have grandparents with so many different nationalities you can't tell what they are," Adair said.

Almost half the people who live in Hawaii today are of "mixed" ancestry. What it means to be mixed is not at all obvious genetically, but for official purposes it means that a person's ancestors fall into more than one of the four "racial" categories identified on U.S. census forms: black, white, Native American, and Asian or Pacific Islander. Intermarriage is a cumulative process, so once an individual of mixed ancestry is born, all of that person's descendants also will be mixed. As intermarriage continues in Hawaii — and already almost half of all marriages are between couples of different or mixed ethnicities — the number of people who will be able to call themselves pure Japanese, or pure Hawaiian, or pure white (*haole* in Hawaiian) will steadily decline.

Hawaii's high rates of intermarriage have fascinated academics for decades. The University of Hawaii sociologist Romanzo Adams wrote an article titled "Hawai'i as a Racial Melting Pot" in 1926, and many scholars since then have extolled Hawaii as a model of ethnic and racial harmony. The researchers have always been a bit vague about the reasons for all this intermarriage; explanations have ranged from the benign climate to the "aloha spirit" of the Native Hawaiians. But their lack of analytic rigor hasn't damped their enthusiasm. One of the goals of the former Center for Research on Ethnic Relations at the University of Hawaii was "to determine why ethnic harmony exists in Hawai'i" and "to export principles of ethnic harmony to the mainland and the world."

The rest of the United States has a smaller percentage of mixed marriages than does Hawaii. But given recent trends, one might wonder if the country as a whole is headed down a road Hawaii took long ago. According to the 2000 census, one in twenty children under the age of eighteen in the United States is mixed, in that their parents fall into more than one racial category. Between the 1990 and 2000 censuses, the number of interracial couples quadrupled. This number — about 1.5 million of 55 million married couples — is not yet high, but because of kinship ties, American families are already much more mixed than they look. Demographer Joshua Goldstein of Princeton University has calculated that about 20 percent of Americans are already in extended families with someone from a different racial group — that is, they or their parents, uncles and aunts, siblings, or children have married someone classified as a member of a different race.

The rapid growth of interracial marriages in the United States and elsewhere marks a new phase in the genetic history of humanity. Since the appearance of modern humans in Africa more than 100,000 years ago, human groups have differentiated in appearance as they have expanded across the globe and have undergone some measure of reproductive isolation. This differentiation has always been limited by the recentness of our common ancestry and by the powerful tendency of groups to mix over time. Still, many human populations have remained sufficiently separate to develop and retain the distinctive physical characteristics we recognize today.

In Hawaii this process is occurring in reverse. It's as if a videotape of our species' history were being played backward at a fantastically rapid speed. Physical distinctions that took thousands of generations to produce are being wiped clean with a few generations of intermarriage.

The vision of the future conjured up by intermarriage in Hawaii can be seductive. When everyone is marrying everyone else, when the ethnic affiliation of most people can no longer be ascertained at a glance, one imagines that ethnic and racial tensions would diminish. But spending some time in Hawaii shows that the future will not be that simple. Despite the high rate of intermarriage here, ethnic and racial tensions haven't really disappeared. They have changed into something else, something less threatening, perhaps, but still divisive. Hawaii may well be a harbinger of a racially mixed future. But it won't be the future many people expect.

Many of the harshest conflicts in the world today are between people who are physically indistinguishable. If someone took a roomful of Palestinians and Israelis from the Middle East, or of Serbs and Albanians from the Balkans, or of Catholics and Protestants from Ireland, or of Muslims and Hindus from northern India, or of Dayaks and Madurese from Indonesia, gave them all identical outfits and haircuts, and forbade them to speak or gesture, no one could distinguish the members of the other group — at least not to the point of being willing to shoot them. The antagonists in these conflicts have different ethnicities, but they have been so closely linked biologically throughout history that they have not developed marked physical differences.

Yet one of the most perverse dimensions of ethnic thinking is the "racialization" of culture — the tendency to think of another people as not just culturally but genetically distinct. In the Yugoslavian war, the Croats caricatured their Serbian opponents as tall and blond, while the Serbs disparaged the darker hair and skin of the Croats — even though these traits are thoroughly intermixed between the two groups. During World War II the countries of Europe fiercely stereotyped the physical attributes of their enemies, despite a history of intermarriage and migration that has scrambled physical characteristics throughout the continent. In Africa the warring Tutsis and Hutus often call attention to the physical differences of their antagonists, but most observers have trouble distinguishing individual members of the two groups solely on the basis of appearance.

The flip side of this biological stereotyping is the elevation of one's own ancestry. The Nazis were the most notorious believers in the purity of their past, but many other groups have similar beliefs. They proclaim themselves to be descended from ancient tribes of noble warriors, or from prominent families in the distant past, or even from famous individuals.

Genetics research has revealed the flaw inherent in any such belief. Every group is a mixture of many previous groups, a fleeting collection of genetic variants drawn from a shared genetic legacy. The Polynesian colonizers of the Hawaiian archipelago are a good example. In 1795 the German anatomist J. F. Blumenbach proposed that the "Malays" — a collection of peoples, including the Polynesians, from southeastern Asia and Oceania — were one of the five races of humanity, in addition to Africans, Caucasians, Mongoloids, and Native Americans. But all of these groups (to the extent that they can be defined) are genetic composites of previous groups. In the case of the Polynesians, this mixing was part of the spread of humans into the Pacific. The last major part of the world to be occupied by humans was Remote Oceania, the widely separated islands scattered in a broad crescent from Hawaii to New Zealand. Before that, humans had been living only in Near Oceania, which includes Australia, Papua New Guinea, and the Bismarck Archipelago. The humans who settled these regions

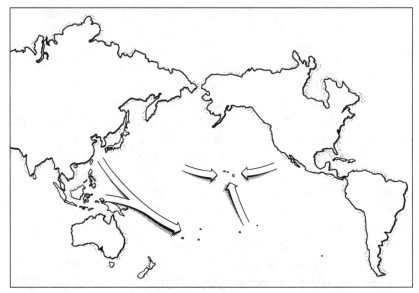

The Polynesian inhabitants of Hawaii are descended from people who lived both in southeastern Asia and in Melanesia, which includes New Guinea and nearby islands. The Polynesians migrated first to the South Pacific islands of Fiji, Tonga, and Samoa, with later migrations taking their descendants from the Marquesas Islands to Hawaii. More recent migrants have included people from Asia, the Americas, and Europe.

were adept at short ocean crossings, but they never developed the kinds of boats or navigation skills needed to sail hundreds of miles to Fiji, Samoa, and beyond.

Then, about 6,000 years ago, rice and millet agriculture made the leap across the Formosa Strait from the mainland of southeastern Asia to Taiwan. From there, agriculture began to spread, island by island, to the south and southeast. With it came two important cultural innovations. The first was the Austronesian language family, which eventually spread halfway around the world, from Madagascar to Easter Island. The second was a suite of new technologies — pottery, woodworking implements, and eventually the outrigger canoe and ways of using the stars to navigate across large expanses of open water. Archaeological evidence shows that people first reached the previously uninhabited island of Fiji about 3,000 years ago. They sailed to Easter Island, their farthest point east, in about A.D. 300 and to New Zealand, their farthest point south, in about 800.

One hypothesis, known as the express-train model of Polynesian

origins, claims that both the knowledge of agriculture and Austronesian languages were carried into the Pacific by people descended almost exclusively from the first farmers who set sail from Taiwan. But
genetic studies have revealed a much more complex picture. Mitochondrial and Y-chromosome haplotypes among today's Polynesians
show that there was extensive mixing of peoples in Near Oceania,
which eventually produced the groups that set sail for the remote islands. Though many of the mitochondrial haplotypes and Y chromosomes of the Polynesians do seem to have come from the mainland of
southeastern Asia and Taiwan, others originated in New Guinea and
its nearby islands — a geographic region known as Melanesia (named
for the generally dark skin of its inhabitants). Geneticists Manfred
Kayser and Mark Stoneking of the Max Planck Institute for Evolutionary Anthropology in Leipzig have dubbed the resulting synthesis the
"slow-boat model." According to this model, today's Polynesians can
trace their ancestry both to the Austronesian speakers who moved
out of southeastern Asia and to the people who already occupied Melanesia.

The Polynesians first reached the Hawaiian Islands around A.D. 400,
probably in a migration from the Marquesas Islands. A subsequent
wave of people migrated to Hawaii from Tahiti between the twelfth
and the fourteenth century. Then the islands saw no more newcomers
until Captain Cook's arrival four centuries later.

The discovery of Hawaii by Europeans did not result in an immediate influx of colonists. The early decades of the nineteenth century
brought just a trickle of settlers to the islands — washed-up sailors, re-
tired captains, British and Russian traders, missionaries. Large-scale
migration began only after the first sugar plantations were established
around the middle of the century. In 1852, three hundred Chinese
men arrived to work the plantations. Over the next century nearly half
a million more workers followed. They came from China, Japan, Ko-
rea, Puerto Rico, Spain, Poland, Austria, Germany, Norway, and Russia. Some of these groups have long since disappeared, blending into
the genetic background. Others still have a significant ethnic presence
on the islands.

A few miles from the Honolulu airport is a vivid reminder of those
times. Hawaii's Plantation Village is one of the few tourist attractions
designed as much for the locals as for mainlanders. It meticulously
recreates a camp town of the type that once dotted the islands, hous-

ing the workers who toiled each day in the sugar and pineapple fields. Each house along the main avenue reflects the ethnicity of the workers who lived there: a large bread oven sits next to the Portuguese house, rice cookers dominate the kitchen of the Chinese house, crucifixes adorn the walls of the Puerto Rican house. A Japanese shrine is a few doors away from the Chinese society building. Down the hill by the taro fields is a *dohyo,* a sumo ring, where the workers wrestled every Sunday afternoon.

Mike Hama showed me around the day I was there. The descendant of Japanese, German, Hawaiian, and Irish grandparents, Hama grew up on a plantation camp in the 1940s. "Kids of all different nationalities played together in these camps," he told me. "We didn't know we were different." They communicated using a pidgin that combined words from many languages. The German kids taught the other kids to polka in the camp social halls. The Japanese kids taught their friends sumo wrestling. When the Japanese emperor visited Hawaii after World War II, according to a widely told if hard-to-verify story, he was so impressed to see wrestlers of all different nationalities in the *dohyo* that when he returned to Japan he opened the country's sumo rings to foreigners.

When Hama was eighteen, he joined the military and was stationed in California. "That was a real awakening for me," he recalls. "For the first time I saw the bigotry that was going on outside Hawaii." He moved back to Hawaii as soon as he could and married a woman of mixed ancestry. His four daughters think of themselves as nothing other than local Hawaiians.

The camp towns disappeared decades ago in Hawaii, yet they have left a remarkable legacy. Large-scale segregation in housing remains rare on the islands. People of all ethnic backgrounds live side by side, just as they did in the camp towns. The only people who live in ghettos are the soldiers on military bases and wealthy haoles who wall themselves off in gated communities. Because neighborhoods are integrated in Hawaii, so are most of the schools. Children of different ethnicities continue to grow up together and marry, just as they did in the camps.

Integrated neighborhoods, integrated schools, high rates of intermarriage — the islands sound as if they should be a racial paradise. But there's actually a fair amount of prejudice here. It pops up in novels, politics, the spiels of standup comics. And it's especially prominent in everyday conversation — "talk stink" is the pidgin term for disrespecting another group.

Some of the prejudice is directed toward haoles, who continue to occupy many of the positions of social and economic prominence on the islands (though their days as plantation overlords are long gone). Nonwhites label haoles as cold, self-serving, arrogant, meddling, loud, and even that old stereotype — smelly (because, it is held, they still do not bathe every day). White kids say they'll get beat up if they venture onto certain nonwhite beaches. Occasionally a rumor sweeps through a school about an upcoming "Kill a Haole" day. The rumors are a joke meant to shock the prevailing sensibilities. But one would not expect such a joke where racial tensions are low.

Other groups come in for similarly rough treatment. The Japanese are derided as clannish and power-hungry, the Filipinos as ignorant and underhanded, the Hawaiians as fat, lazy, and fun-loving. And, as is true of stereotypes everywhere, the objects of them have a tendency to reinforce them, either by too vigorously denying or too easily repeating them.

"Intermarriage may indicate tolerance," says Jonathan Okamura, an anthropologist at the University of Hawaii, "but it doesn't mean we have an egalitarian society on a larger scale." Though he calls his viewpoint a "minority position," Okamura holds that racial and ethnic prejudice is deeply ingrained in the institutional structures of everyday life in Hawaii. For example, the integration of the public schools is deceptive, he says. Well-off haoles, Chinese, and Japanese send their children to private schools, and the public schools are underfunded. "We've created a two-tiered system that makes inequality increasingly worse rather than better," says Okamura. Meanwhile the rapid growth of the tourism industry in Hawaii has shut off many traditional routes to economic betterment. Tourism produces mostly low-paying jobs in sales, service, and construction, Okamura points out, so people have few opportunities to move up career ladders.

Of course, talented and lucky individuals still get ahead. "Students with parents who didn't go to college come to the university and do well — that happens all the time," Okamura says. "But it doesn't happen enough to advance socioeconomically disadvantaged groups in society."

Several ethnic groups occupy the lower end of the socioeconomic scale, but one in particular stands out: the people descended from the island's original inhabitants. Native Hawaiians have the lowest incomes and highest unemployment rates of any ethnic group. They have the most health problems and the shortest life expectancy. They are the least likely to go to college and the most likely to be incarcerated.

Then again, applying statistics like these to a group as large and diverse as Native Hawaiians is inevitably misleading. Individuals with some Hawaiian ancestry make up a fifth of the population in Hawaii. Some are successful; some are not. Some are consumed by native issues; others pay them no mind. And Native Hawaiians are much less marginalized in Hawaii than are, for example, Native Americans in the rest of the United States. Hawaiian words, names, and outlooks have seeped into everyday life on the islands, producing a cultural amalgam that is one of the state's distinct attractions.

Native Hawaiians should not be seen as simply another ethnic group, the leaders of their community point out. Other cultures have roots elsewhere; people of Japanese, German, or Samoan ancestry can draw from the traditions of an ancestral homeland to sustain an ethnic heritage. If the culture of the Native Hawaiians disappears, it will be gone forever. Greater recognition of the value and fragility of this culture has led to a resurgence of interest in the Hawaiian past. Schools with Hawaiian language immersion programs have sprung up around the islands to supplement the English that children speak at home. Traditional forms of Hawaiian dance, music, canoeing, and religion all have undergone revivals.

This Hawaiian Renaissance also has had a political dimension. For the past several decades a sovereignty movement has been building among Native Hawaiians that seeks some measure of political autonomy and control over the lands that the U.S. government seized from the Hawaiian monarchy at the end of the nineteenth century. Reflecting the diversity of the native population, several sovereignty organizations have carried out a sometimes unseemly struggle over strategies and goals. One radical faction advocates the complete independence of the islands from the United States. More moderate groups have called for the establishment of a Native Hawaiian nation modeled on the Indian tribes on the mainland. Native Hawaiians would have their own government, but it would operate within existing federal and state frameworks, and its citizens would remain Americans.

Native Hawaiian sovereignty faces many hurdles, and it is premature to harp on exactly how it would work. But whenever the topic comes up in discussion, a question quickly surfaces: exactly who is a Native Hawaiian? "Pure" Hawaiians with no non-Hawaiian ancestors probably number just a few thousand. Many Native Hawaiians undoubtedly have a preponderance of Hawaiian ancestors, but no clear line separates natives from nonnatives. Some people who call themselves Native Hawaiians probably have little DNA from Polynesian ancestors.

Past legislation has waffled on this issue. Some laws define Native Hawaiians as people who can trace at least half their ancestry to people living in Hawaii before the arrival of Captain Cook. Others define as Hawaiian anyone who has even a single precontact Hawaiian ancestor. These distinctions are highly contentious for political and economic as well as cultural reasons. Many state laws restrict housing subsidies, scholarships, economic development grants, and other benefits specifically to Native Hawaiians.

As the study of genetics and history has progressed, an obvious idea has arisen. Maybe science could resolve the issue. Maybe a genetic marker could be found that occurs only in people descended from the aboriginal inhabitants of Hawaii. Then anyone with that marker could be considered a Native Hawaiian.

No one is better qualified to judge this idea than Rebecca Cann, a professor of genetics at the University of Hawaii. Cann was the young graduate student at the University of California at Berkeley who, with Mark Stoneking, did much of the work that led to the unveiling of mitochondrial Eve. She haunted hospital delivery rooms to obtain mitochondrion-rich placentas, which at that time was the only way to get enough mitochondrial DNA to sequence. She ran gels and compared nucleotides. Her faculty adviser, Allan Wilson, landed mitochondrial Eve on the cover of *Newsweek*, but Cann did the footwork.

She moved to Hawaii even before mitochondrial Eve made headlines, responding to an ad in *Science* magazine for a job. She's been here ever since, though her flat American accent still betrays a childhood spent in Iowa. She met me at the door of her office, in the foothills above Honolulu, dressed in sandals and a patterned Hawaiian dress. "I think we correctly anticipated many of the applications and potential problems of this research," she said, "right down to people wanting to clone Elvis from a handkerchief he'd used to wipe his brow.

What we didn't understand was the degree to which religious and cultural beliefs would dictate attitudes toward genetic materials. In Hawaii, for instance, there's a very strong belief in *mana*, in the power of the spirit, which is contained in the remains of a person's ancestors. The absolute disgust that many people have toward the desecration of a grave — that was a cultural eye-opener for me."

Despite the occasional cultural difficulties, Cann has continued her study of human genetics in Hawaii and has played an important role in piecing together the prehistory of the Pacific. By comparing the mitochondrial DNA sequences of people on various islands, she has traced the gradual eastward spread of modern humans from southeastern Asia and Melanesia. She has discovered that men and women had different migration patterns into the Pacific and has even detected tantalizing evidence, still unconfirmed, of genetic contacts between Pacific Islanders and South Americans. "I'm convinced that our history is written in our DNA," she told me.

Yet she cautioned against using genetics to determine ethnicity. "I get people coming up to me all the time and saying, 'Can you prove that I'm a Hawaiian?'" She can't, she said, at least not with a high degree of certainty. A given individual might have a mitochondrial haplotype that is more common among Native Hawaiians. But the ancestors of the aboriginal Hawaiians also gave rise to other Pacific populations, so a mitochondrial sequence characteristic of Native Hawaiians could have come from a Samoan or Filipino ancestor.

Also, a person's mitochondrial DNA is not necessarily an accurate indication of ancestry. The only way for a person to have mitochondrial DNA from a woman who lived in Hawaii before the arrival of Captain Cook is for that person to have an unbroken line of grandmothers dating back to that woman. But because groups have mixed so much in Hawaii, mitochondrial lineages have become thoroughly tangled. People who think of themselves as Native Hawaiian could easily have had non-Hawaiian female ancestors sometime in the past eleven generations, which would have given them mitochondrial DNA from another part of the world.

These genetic exchanges are also common elsewhere in the world, even in populations that think of themselves as less mixed. Most native Europeans, for example, have mitochondrial DNA characteristic of that part of the world. But some have mitochondrial DNA from elsewhere — southern Africa, or eastern Asia, or even Polynesia

— brought to Europe over the millennia by female immigrants. The British matron who has a mitochondrial haplotype found most often in southern Africans is not an African, just as the Native Hawaiian with mitochondrial DNA from a German great-great-grandmother does not automatically become German.

This confusion of genetic and cultural identities becomes even greater with the Y chromosome, given the ease with which that chromosome can insert itself into a genealogy. Most of the early migrants to Hawaii, for example, were males, especially the plantation workers. Those males mated with native women more often than native men mated with immigrant women, so nonnative Y chromosomes are now more common in mixed populations than nonnative mitochondrial DNA. In some populations in South America, virtually all the Y chromosomes are from Europe and all the mitochondrial DNA is from indigenous groups.

The mixing of genes can cause great consternation, but it is the inevitable consequence of our genetic history. Several years ago a geneticist in Washington, D.C., began offering to identify the homelands of the mitochondrial DNA and Y chromosomes of African Americans. The service foundered for several reasons, but one was that 30 percent of the Y chromosomes in African-American males come from European ancestors.

Within a few years geneticists will be able to use DNA sequences from all the chromosomes to trace ancestry. But these histories will be just as convoluted as those of mitochondrial DNA and the Y. Granted, geneticists will be able to make statistical assessments. They will be able to say, for instance, that a given person has such and such a probability of descent from a Native Hawaiian population, and in some cases the probability will be very high. But probabilities don't convey the cold, hard certainties that people want in their genealogies.

Beyond the purely genetic considerations are the social ones. When children are adopted from one group into another, they become a member of that group socially, yet their haplotypes and those of their descendants can differ from the group norm. Rape is another way in which the genetic variants of groups mix. And sometimes people from one group make a conscious decision to join another and are gladly accepted, despite their different genetic histories.

"I get nervous when people start talking about using genetic markers to prove ethnicity," Cann told me. "I don't believe that biology is destiny. Allowing yourself to be defined personally by whatever your DNA sequence is, that's insane. But that's exactly what some people are going to be tempted to do."

When geneticists look at our DNA, they do not see a world of rigidly divided groups each going its own way. They see something much more fluid and ambiguous — something more like the social structures that have emerged in Hawaii as intermarriage has accelerated.

The most remarkable aspect of ethnicity in Hawaii is its loose relation to biology. Many people have considerable latitude in choosing their ethnic affiliations. Those of mixed ancestry can associate with the ethnicity of a parent, a grandparent, or a more distant ancestor. They can partition their ethnic affiliations: they can be Chinese with their Chinese relatives; Native Hawaiian with their native kin; and just plain local with their buddies. The community of descent that a person associates with has become more like a professional or religious affiliation, a connection over which a person has some measure of control.

People whose ancestors are from a single ethnic group have fewer options, but they, too, can partake of at least some of Hawaii's ethnic flexibility. Young whites, for example, sometimes try to pass themselves off as mixed by maintaining an especially dark tan. Among many young people, dating someone from a different ethnic group is a social asset rather than a liability, in part because of the doors it opens to other communities. Many prospective students at the University of Hawaii simply mark "mixed" in describing their ethnicity on application forms, even if both parents have the same ethnic background. "My students say they don't want to be pigeonholed," says Okamura. "That way they can identify with different groups."

Hawaii's high rates of intermarriage also contribute greatly to the islands' ethnic flux. Ethnicity is not defined just by who one's ancestors were. It also is defined prospectively — by the group into which one is expected to marry. For most young people in Hawaii, the pool of marriageable partners encompasses the entire population. Relations among groups are inevitably less fractious when their members view each other as potential mates.

Of course, ethnic and even "racial" groups still exist in Hawaii, and they will for a long time. Despite the rapid growth of intermarriage in Hawaii and elsewhere, the mixing of peoples takes generations, not a few years or even decades. Most people around the world still choose marriage partners who would be classified as members of the same "race." In many parts of the world — the American Midwest, China, Iceland — few other options are available. Five hundred years from now, unless human societies undergo drastic changes, Asians, Africans, and Europeans still will be physically distinguishable.

But the social effects of intermarriage are much more immediate than are the biological effects. Socially, intermarriage can quickly undermine the idea that culture has biological roots. When a substantial number of mixed individuals demonstrate, by their very existence, that choices are possible, that biology is not destiny, the barriers between groups become more permeable. Ethnicity in Hawaii, for example, seems far less stark and categorical than it does in the rest of the United States. The people of Hawaii recognize overlaps and exceptions. They are more willing to accept the haole who claims to have non-European ancestors or the Native Hawaiian who affiliates with Filipinos. It's true that people talk about the differences among groups all the time, but even talking about these differences, rather than rigidly ignoring them, makes them seem less daunting. Expressions of social prejudice in Hawaii are more like a form of social banter, like a husband and wife picking at each other's faults.

The logical endpoint of this perspective is a world in which people are free to choose their ethnicity regardless of their ancestry. Ethnicity is not yet *entirely* voluntary in Hawaii, but in many respects the islands are headed in that direction. State law, for example, is gradually coming to define a Native Hawaiian as anyone with a single Hawaiian ancestor. But at that point ethnicity becomes untethered from biology — it is instead a cultural, political, or historical distinction. People are no longer who they say they are because of some mysterious biological essence. They have chosen the group with which they want to affiliate.

Genetically, this view of ethnicity makes perfect sense. Our DNA is too tightly interconnected to use biology to justify what are essentially social distinctions. Our preferences, character, and abilities are not determined by the biological history of our ancestors. They depend on our individual attributes, experiences, and choices. As this inescapable conclusion becomes more widely held, our genetic histories inevitably will become less and less important. When we look at another person, we won't think Asian, black, or white. We'll just think: person.

*

In his novel *Siddhartha*, Hermann Hesse tells the story of a young man in ancient India, a disciple of an inspired teacher, who sets out to find the reality beneath the world of appearance. After years of study and wandering, Siddhartha becomes a ferryman, learning from his predecessor how to listen to the voices in the passing river. One day a childhood friend named Govinda comes to the river. Siddhartha and Govinda have a long conversation about the interdependence of illusion and truth, about the existence of the past and future in the present, about the need not just to think about the world but to love it. Finally Govinda asks Siddhartha how he has achieved such peace in his life. Siddhartha replies, "Kiss me on the forehead, Govinda." Govinda is surprised by the request, but out of respect for his friend he complies. When he touches Siddhartha's forehead with his lips, he has a wondrous vision:

> He no longer saw the face of his friend Siddhartha. Instead he saw other faces, many faces, a long series, a continuous stream of faces — hundreds, thousands, which all came and disappeared and yet all seemed to be there at the same time, which all continually changed and renewed themselves and which were all yet Siddhartha. . . . He saw the face of a newly born child, red and full of wrinkles, ready to cry. He saw the face of a murderer, . . . He saw the naked bodies of men and women in the postures and transports of passionate love. . . . Each one was mortal, a passionate, painful example of all that is transitory. Yet none of them died, they only changed, were always reborn, continually had a new face; only time stood between one face and another.

I began this book by calling attention to the different appearances of human beings. I conclude it now by calling attention to the opposite. Throughout human history, groups have wondered how they are related to one another. The study of genetics has now revealed that we all are linked: the Bushmen hunting antelope, the mixed-race people of South Africa, the African Americans descended from slaves, the Samaritans on their mountain stronghold, the Jewish populations scattered around the world, the Han Chinese a billion strong, the descendants of European settlers who colonized the New World, the Native Hawaiians who look to a cherished past. We are members of a single human family, the products of genetic necessity and chance, borne ceaselessly into an unknown future.

NOTES

ACKNOWLEDGMENTS

INDEX

Notes

Introduction: The Human Pageant

page

1   *Human beings really are one of the most physically varied species on Earth.*
Vincent Sarich makes this point in "The Final Taboo: Race Differences in
Ability," *Skeptic* 8, no. 1 (2000):38–43.

1   *almost all the armed conflicts in the world take place not between nations but
between groups . . .* For an overview of recent conflicts, see chapter 1 of *Ethnicity and Race: Making Identities in a Changing World* by Stephen Cornell
and Douglas Hartmann (Thousand Oaks, Calif.: Pine Forge Press, 1998).

2   *In the United States, more than three-quarters of African Americans and European Americans alike judge relations . . . to be only fair or poor.* These poll results appear in "Three Is Not Enough" by Sharon Begley, in *The Biological
Basis of Human Behavior: A Critical Review,* edited by Robert Sussman (Upper Saddle River, N.J.: Prentice Hall, 1999).

2   *"The problem of the twentieth century is the problem of the color line."* Du Bois
first made this remark at a London conference in 1900. It also appears in his
1903 collection of essays *The Souls of Black Folk* (New York: W. W. Norton,
1999).

2   *More than 100,000 years ago, a small group of humans lived in this region.* Jon-
athan Kingdon sketches a compelling portrait of the life of our ancestors in
Self-Made Man: Human Evolution from Eden to Extinction? (New York: John
Wiley and Sons, 1993).

2   *These people would never have passed what anthropologists call the subway test.*
The test was first proposed in a paper by William Straus and A. J. E. Cave,
"Pathology and Posture of Neanderthal Man," *Quarterly Review of Biology* 32
(1957):348–63.

3   *More than 1 million archaic humans may have been scattered across the Old
World . . .* This estimate is taken from Kenneth Weiss, "On the Number of
Members of the Genus *Homo* Who Have Ever Lived, and Some Evolutionary
Implications," *Human Biology* 56 (1984):637–49.

3   *Every single one of the 6 billion people on the planet today is descended from the*

small group of anatomically modern humans who once lived in eastern Africa. Though this conclusion has generated ferocious arguments over the past two decades, the vast majority of geneticists and most paleoanthropologists now accept it with few reservations. See, for example, "Using Mitochondrial and Nuclear DNA Markers to Reconstruct Human Evolution" by Lynn Jorde, Michael Bamshad, and Alan Rogers, *BioEssays* 20 (1998):126–36.

3 *the two groups interbred very little, if at all.* Many geneticists have looked, without success, for evidence of such mixing. See, for example, "Short Tandem-Repeat Polymorphism/*Alu* Haplotype Variation at the PLAT Locus: Implications for Modern Human Origins" by Sarah Tishkoff and eleven colleagues, *American Journal of Human Genetics* 67 (2000):901–25.

3 *The cave paintings of Europe . . .* In *A History of Warfare* (New York: Alfred A. Knopf, 1993), John Keegan notes that warfare did not become an obvious subject of art until after the invention of agriculture, tens of thousands of years after modern humans moved into Europe.

4 *of the several billion modern humans who lived before the invention of agriculture . . .* Chapter 5 includes estimates of the numbers of modern humans who have lived during various periods of history. For an overview of modern human fossils and artifacts, see chapter 7 of *The Human Career: Human Biological and Cultural Origins,* 2nd ed., by Richard Klein (Chicago: University of Chicago Press, 1999).

5 *Research into the genetic causes of disease is simultaneously revealing the histories of groups and of individuals.* Lynn Jorde, W. Scott Watkins, and Michael Bamshad make this point in "Population Genomics: A Bridge from Evolutionary History to Genetic Medicine," *Human Molecular Genetics* 10 (2001):2199–2207.

6 *The study of human genetic differences is one of the most contentious areas of modern science.* A good description of current controversies in human genetics is *Unnatural Selection: The Promise and the Power of Human Gene Research* by Lois Wingerson (New York: Bantam Books, 1998).

6 *we have time to think about the dilemmas that those applications will pose.* Allen Buchanan, Dan Brock, Norman Daniels, and Daniel Wikler analyze many of these dilemmas in their book *From Chance to Choice: Genetics and Justice* (New York: Cambridge University Press, 2000).

1. The End of Evolution

11 *The lives of . . . the !Kung San . . . have changed dramatically in recent decades.* Richard Lee describes the lives of the Bushmen in a village a few miles away from the one I visited in *The Dobe Ju/'hoansi,* 2nd ed. (New York: Harcourt Brace, 1993).

12 *The Bushmen are the original people of southern Africa.* Hilary Deacon discusses the ancient history of the Bushmen in "Southern Africa and Modern Human Origins," *Philosophical Transactions of the Royal Society of London B* 337 (1992):177–83. See also his book, written with Janette Deacon, *Human Beginnings in South Africa: Uncovering the Secrets of the Stone Age* (Walnut Creek, Calif.: Altamira Press, 1999).

13 *over the last few millennia, other groups have steadily encroached on their homelands.* For a succinct description of the interactions among groups in southern Africa, see *A History of South Africa*, rev. ed., by Leonard Thompson (New Haven: Yale University Press, 1995).

13 *A late-nineteenth-century tally . . .* Pippa Skotnes includes this record in *Miscast: Negotiating the Presence of the Bushmen* (Athens: Ohio University Press, 1998).

13 *Well into the twentieth century, anthropologists were speculating . . .* Milford Wolpoff and Rachel Caspari describe the sorry history of speculation about groups of people being different species in *Race and Human Evolution* (New York: Simon and Schuster, 1997).

15 *these banding patterns are essentially the same for people anywhere in the world.* Elizabeth Nickerson and David Nelson discuss the chromosomal differences among humans, chimps, gorillas, and orangutans in "Molecular Definition of Pericentric Inversion Breakpoints Occurring During the Evolution of Humans and Chimpanzees," *Genomics* 50 (1998):368–72. For a good sense of how the chromosomes of humans, chimps, gorillas, and orangutans differ, see Jorge Yunis and Orn Prakash, "The Origin of Man: A Chromosomal Pictorial Legacy," *Science* 215 (1982):1525–30.

18 *The four most important events in human evolution . . .* William Howells discusses all four in *Getting Here: The Story of Human Evolution* (Washington, D.C.: Compass Press, 1997). A good review of the evolution of our genus is "*Australopithecus* to *Homo*: Transformations in Body and Mind" by Henry McHenry and Katherine Coffing, *Annual Review of Anthropology* 29 (2000):125–46.

20 *we are the product of a relentless winnowing process . . .* Paleoanthropologists have argued endlessly over how to divide fossil members of the genus *Homo* into species. In general, I have taken the position of a "splitter" rather than a "lumper" in this chapter, because a diversity of species accords well with the genetic evidence. For a good summary of the paleoanthropological arguments, see "The Changing Face of Genus *Homo*" by Bernard Wood and Mark Collard, *Evolutionary Anthropology* 8 (1999):195–207.

20 *Today, just a single human species lives on the earth.* Ian Tattersall, an ardent splitter, presents his view of human evolution in *Becoming Human: Evolution and Human Uniqueness* (New York: Harcourt Brace, 1998). A dramatically illustrated article by Tattersall that describes the plethora of human and prehuman species is "Once We Were Not Alone," *Scientific American* (Jan. 2000):56–62.

20 *Groves has put human evolution in context.* Colin Groves analyzes human fossils and evolution in *A Theory of Human and Primate Evolution*, 2nd ed. (Oxford, Eng.: Oxford University Press, 1991). For a more popular treatment of the fossil and genetic evidence surrounding the origins of our species, see his article "The Origin of Modern Humans" in *Interdisciplinary Science Reviews* 29 (1994):23–34. Groves's views on the formation of new species can be found in "How Old Are Subspecies? A Tiger's Eye-View of Human Evolution," *Archaeology in Oceania* 27 (1992):153–60.

22 *Dense populations produce rich stores of genetic diversity on which evolution*

can act. An interesting article on speciation processes in the center of a species' range is "Genetics and Speciation" by Jerry Coyne, *Nature* 355 (1992):511–15.

22   *The highlands of Kenya and Tanzania are a biological paradise.* Jonathan Kingdon describes modern humans' Eden in chapter 2 of *Self-Made Man: Human Evolution from Eden to Extinction?* (New York: John Wiley and Sons, 1993).

23   *this group must have been very small, occupying an area that may have been no larger than present-day Israel.* Henry Harpending and five other researchers speculate about the geographic distribution of our ancestral population in "Genetic Traces of Ancient Demography," *Proceedings of the National Academy of Sciences* 95 (1998):1961–67.

23   *The oldest known fossil skull thought to belong to a member of this group . . . African Exodus: The Origins of Modern Humanity* by Christopher Stringer and Robin McKie (New York: Henry Holt, 1996) discusses this skull at length.

24   *Some paleoanthropologists, for example, adhere to a much more gradualist view.* For a thorough exposition of this perspective, see *Race and Human Evolution* by Wolpoff and Caspari.

24   *But in 1987 it came under fire from an entirely new quarter.* The paper by Rebecca Cann, Mark Stoneking, and Allan Wilson was "Mitochondrial DNA and Human Evolution," *Nature* 325 (1987):31–36. As is discussed in Chapter 11, several other investigators anticipated their results. For examples, see the articles by Masatoshi Nei and Arun Roychoudhury, "Genic Variation Within and Between the Three Major Races of Man: Caucasoids, Negroids, and Mongoloids," *American Journal of Human Genetics* 26 (1974):421–43; M. J. Johnson and coauthors, "Radiation of Human Mitochondrial DNA Types Analyzed by Restriction Endonuclease Cleavage Patterns," *Journal of Molecular Evolution* 19 (1983):255–71; and J. S. Wainscoat and coauthors, "Evolutionary Relationships of Human Populations from an Analysis of Nuclear DNA Polymorphisms," *Nature* 319 (1986):491–93.

26   *Men pass most of their Y chromosome on to their sons in the same way . . .* Good descriptions of how the Y chromosome can be used in historical studies appear in "The Role of the Y Chromosome in Human Evolutionary Studies" by Michael Hammer and Stephen Zegura, *Evolutionary Anthropology* (1996):116–33; and "Fathers and Sons: The Y Chromosome and Human Evolution" by Mark Jobling and Chris Tyler-Smith, *Trends in Genetics* 11 (1995):449–56.

27   *something like 86,000 individuals . . . are the sources of all the human DNA in existence today.* This estimate comes from Carsten Wiuf and Jotun Hein, "On the Number of Ancestors to a DNA Sequence," *Genetics* 147 (1997):1459–68.

27   *a coalescence is much more likely to occur when a population is small.* Roderic Page and Edward Holmes provide a thorough analysis of coalescence processes and bottlenecks in *Molecular Evolution: A Phylogenetic Approach* (Malden, Mass.: Blackwell, 1998).

29   *In the most thorough computerized analysis of fossilized skulls ever done . . .* Marta Lahr describes her analysis and conclusions in *The Evolution of Mod-*

ern Human Diversity: A Study of Cranial Variation (Cambridge, Eng.: Cambridge University Press, 1996). See particularly part III, "The Evolution of Modern Human Cranial Diversity from a Single Ancestral Source."

29 *With the appearance of modern humans, the large-scale evolution of our species essentially ceased.* Richard Klein makes this point in *The Human Career: Human Biological and Cultural Origins*, 2nd ed. (Chicago: University of Chicago Press, 1999).

29 *The chimpanzees living on a single hillside in Africa have more than twice as much variety . . .* See "Great Ape DNA Sequences Reveal a Reduced Diversity and an Expansion in Humans" by Henrik Kaessmann, Victor Wiebe, Gunter Weiss, and Svante Pääbo, *Nature Genetics* 27 (2001):155–56.

30 *But of all the histories described in this book, that of the Bushmen is the most straightforward.* That genetic history is described in the paper "mtDNA Variation in the South African Kung and Khwe — and Their Genetic Relationships to Other African Populations" by Yu-Sheng Chen and five other researchers, *American Journal of Human Genetics* 66 (2000):1362–83.

30 Map. The possible homeland of modern humans comes from Lahr, *The Evolution of Modern Human Diversity*, p. 271. The prehistoric distribution of the Bushmen is diagrammed by L. Luca Cavalli-Sforza, Paolo Menozzi, and Alberto Piazza in *The History and Geography of Human Genes*, abridged ed. (Princeton: Princeton University Press, 1994), p. 160. The distribution of today's Bushmen is from the entry "Languages of the World" in *Encyclopedia Britannica*.

2. Individuals and Groups

36 *The nucleotide sequences of a child's chromosomal DNA typically differ at about one hundred locations . . .* This estimate comes from "Rates of Spontaneous Mutations" by John Drake, Brian Charlesworth, Deborah Charlesworth, and James Crow, *Genetics* 148 (1998):1667–86.

36 *That's what geneticists are doing when they use genetic variations to reconstruct human history.* Two excellent reviews of anthropological genetics are "The Genetical Archaeology of the Human Genome," by Arndt von Haeseler, Antti Sajantila, and Svante Pääbo, *Nature Genetics* 14 (1995):135–40; and "Human Evolution" by Svante Pääbo, *Trends in Cell Biology* 9, no. 12 (1999):M13–16.

37 *Soodyall's specialty is the mutational history of mitochondrial DNA.* Himla Soodyall and Trefor Jenkins describe their work in "Mitochondrial DNA Polymorphisms in Khoisan Populations from Southern Africa," *Annals of Human Biology* 56 (1992):315–24. For distinctions among African groups, see "African Populations and the Evolution of Human Mitochondrial DNA" by Linda Vigilant and four coauthors, *Science* 253 (1991):1503–7.

38 *A couple of caveats need to be kept in mind when interpreting mitochondrial DNA sequences.* Good discussions of these caveats appear in "Coming to Terms with Human Genetic Variation" by Kenneth Weiss, *Annual Review of Anthropology* 27 (1998):273–300; "Genetics of Modern Human Origins and Diversity" by John Relethford, *Annual Review of Anthropology* 27 (1998):1–

23; and "Population Bottlenecks and Pleistocene Human Evolution" by John Hawks, Keith Hunley, Sang-Hee Lee, and Milford Wolpoff, *Molecular Biology and Evolution* 17 (2000):2–22.

39 *At least 20 percent of all conceptions end in miscarriages.* This statistic comes from chapter 10 of *Concepts of Genetics* by William Klug and Michael Cummings, 6th ed. (Upper Saddle River, N.J.: Prentice Hall, 2000).

40 *When people live in equatorial regions, dark skin is a great advantage.* Stephen Molnar provides a thorough discussion of skin color in chapter 5 of *Human Variation: Races, Types, and Ethnic Groups,* 4th ed. (Upper Saddle River, N.J.: Prentice Hall, 1998). Jared Diamond discusses the many anomalies of human pigmentation in chapter 6 of *The Third Chimpanzee: The Evolution and Future of the Human Animal* (New York: HarperCollins, 1992). The vitamin D hypothesis for the lightening of skin is explored in "The Evolution of Human Skin Coloration" by Nina Jablonski and George Chaplin, *Journal of Human Evolution* 39 (2000):57–106. Recently Rosalind M. Harding and ten other researchers examined the genetics of an important skin-color protein in "Evidence for Variable Selective Pressures at MC1R," *American Journal of Human Genetics* 66 (2000):1351–61. They concluded that in Africa the protein is under strong selective pressure and that any loss of function is harmful, whereas the protein does not seem to be selected for outside of Africa.

43 *every human group . . . is a complex mixture of previous groups.* Richard Lewontin cites the example of the English in chapter 7 of *Human Diversity* (New York: W. H. Freeman, 1995). John Moore analyzes the formation of groups in detail in "Putting Anthropology Back Together Again: The Ethnogenetic Critique of Cladistic Theory," *American Anthropologist* 96 (1994):925–48.

44 *any given person is the descendant of many ancestors.* Susumu Ohno describes the fascinating mathematics of our ancestors in "The Malthusian Parameter of Ascents: What Prevents the Exponential Increase of One's Ancestors," *Proceedings of the National Academy of Sciences* 93 (1996):15276–78.

44 *The actual estimate is around 375 million.* This estimate is from Massimo Livi-Bacci, *A Concise History of World Population,* 2nd ed. (Malden, Mass.: Blackwell, 1997).

45 *when a U.S. president is elected, professional genealogists . . . turn up many circles of inheritance.* Lois Horowitz offers many such examples in *Dozens of Cousins: Blue Genes, Horse Thieves, and Other Relative Surprises in Your Family Tree* (Berkeley: Ten Speed Press, 1999).

46 *Joseph Chang, a statistician at Yale University, has recently shown . . .* Chang's work appears in "Recent Common Ancestors of All Present-Day Individuals," *Advances in Applied Probability* 31 (1999):1002–26. An earlier article that anticipated some of these results is "Ancestors at the Norman Conquest" by Kenneth Wachter in *Genealogical Demography,* edited by Bennett Dyke and Warren Morrill (New York: Academic Press, 1980).

48 *The percentages of mistaken paternity are not nearly so low.* Sally Macintyre and Anne Sooman discuss what they term the "delicate social issue" of biological and nonbiological fathers in "Non-Paternity and Prenatal Genetic Screening," *Lancet* 338 (1991):869–71.

49 *Most human groups are the products of culture, not biology.* Stephen Cornell and Douglas Hartmann provide a thorough discussion of human groups in *Ethnicity and Race: Making Identities in a Changing World* (Thousand Oaks, Calif.: Pine Forge Press, 1998). Another good discussion of these issues can be found in *Ethnic and Racial Consciousness* by Michael Banton, 2nd ed. (New York: Addison Wesley Longman, 1997). *Ethnicity,* edited by John Hutchinson and Anthony Smith (New York: Oxford University Press, 1996), compiles many excellent essays and excerpts on the subject.

50 *Even when this lineage was young, groups of modern humans within Africa had probably begun to diverge.* Two informative though technical articles on the relationships among African populations are "Combined Use of Biallelic and Microsatellite Y-Chromosome Polymorphisms to Infer Affinities among African Populations" by Rosaria Scozzari and sixteen coauthors, *American Journal of Human Genetics* 65 (1999):829–46; and "mtDNA Variation in the South African Kung and Khwe — and Their Genetic Relationships to Other African Populations" by Yu-Sheng Chen and five coauthors, *American Journal of Human Genetics* 66 (2000):1362–83. The mitochondrial relationships between Europeans and Africans are described in "Mitochondrial Footprints of Human Expansions in Africa" by Elizabeth Watson, Peter Forster, Martin Richards, and Hans-Jürgen Bandelt, *American Journal of Human Genetics* 61 (1997):691–704.

50 *The Utah researchers call this hypothesis the "weak Garden of Eden model."* Henry Harpending, Stephen Sherry, Alan Rogers, and Mark Stoneking describe this model in "The Genetic Structure of Ancient Human Populations," *Current Anthropology* 34 (1993):483–96. Marta Lahr presents the fossil, archaeological, and genetic evidence from this period in chapter 11 of *The Evolution of Modern Human Diversity: A Study of Cranial Variation* (Cambridge, Eng.: Cambridge University Press, 1996).

50 *It is tempting to equate these early subdivisions within Africa . . .* For an overview of African prehistory, see *The Peopling of Africa: A Geographic Interpretation* by James Newman (New Haven: Yale University Press, 1995). Jared Diamond discusses the human diversity of Africa in chapter 19 of *Guns, Germs, and Steel: The Fates of Human Societies* (New York: W. W. Norton, 1997). Luca Cavalli-Sforza, Paolo Menozzi, and Alberto Piazza describe the genetics and history of human groups in Africa in chapter 3 of *The History and Geography of Human Genes,* abridged ed. (Princeton: Princeton University Press, 1994). For a contrasting view, see "Genes, Tribes, and African History" by Scott MacEachern, *Current Anthropology* 41 (June 2000):357–84.

51 *Also, all three groups have expanded and contracted over time.* Grahame Clark summarizes the archaeology of African prehistory in chapter 5 of *World Prehistory in New Perspective,* 3rd ed. (Cambridge, Eng.: Cambridge University Press, 1977). John Reader brings a broader perspective to the same time period in chapters 14 and 15 of *Africa: A Biography of the Continent* (New York: Alfred A. Knopf, 1998). See also *African Archaeology,* 2nd ed., by David Phillipson (Cambridge, Eng.: Cambridge University Press, 1993).

52 Map. The distribution of population groups within Africa comes from Colin McEvedy, *The Penguin Atlas of African History,* rev. ed. (New York: Penguin,

1995), p. 20. The initial stages of the Bantu expansion are described on pages 32–34 of that book.

52 *Initially, the Bantu speakers spread east* . . . Luca Cavalli-Sforza and Francesco Cavalli-Sforza describe this and many other prehistoric migrations in *The Great Human Diasporas: The History of Diversity and Evolution* (Reading, Mass.: Perseus Books, 1995). For a general description of the Bantu speakers, see *The People of Africa* by Jean Hiernaux (New York: Charles Scribner's Sons, 1975).

53 *The DNA of modern Africans clearly reflects this Bantu expansion.* Himla Soodyall, Linda Vigilant, Adrian Hill, Mark Stoneking, and Trefor Jenkins describe the nine-base-pair deletion in "mtDNA Control-Region Sequence Variation Suggests Multiple Independent Origins of an 'Asian-Specific' 9-bp Deletion in Sub-Saharan Africans," *American Journal of Human Genetics* 58 (1996):595–608.

3. The African Diaspora and the Genetic Unity of Modern Humans

54 *The city where I live* . . . *Urban Odyssey: A Multicultural History of Washington, D.C.*, edited by Francine Curro Cary (Washington, D.C.: Smithsonian Institution Press, 1996), contains historical essays describing some of the many different ethnic groups in the city.

55 *But almost all African Americans also have a large number of European-American ancestors.* Esteban Parra and ten colleagues calculate the average number of European ancestors in "Estimating African-American Admixture Proportions by Use of Population-Specific Alleles," *American Journal of Human Genetics* 63 (1998):1839–51. See also "Caucasian Genes in American Blacks: New Data" by Ranajit Chakraborty, Mohammad Kamboh, M. Nwankwo, and Robert E. Ferrell, *American Journal of Human Genetics* 50 (1992):145–55, and the response by T. Edward Reed in *American Journal of Human Genetics* 51 (1992):678–79.

55 *many European Americans have relatively recent African-American ancestors.* Brent Staples discusses the African-American ancestors of European Americans in "The Real American Love Story: Why America Is a Lot Less White Than It Looks," *Slate* (Oct. 4, 1999).

55 *These "passers"* . . . *have been a powerful force for demographic mixing in the United States.* G. Reginald Daniel assesses that influence in "Passers and Pluralists: Subverting the Racial Divide" in *Racially Mixed People in America*, edited by Maria Root (Newbury Park, Calif.: Sage Publications, 1992).

55 *Several of the children of Thomas Jefferson and his African-American slave Sally Hemings* . . . A wonderful book that traces the history of Jefferson's descendants on both sides of the family is *Jefferson's Children: The Story of an American Family* by Shannon Lanier and Jane Feldman (New York: Random House, 2000).

57 *Scholars have cited many reasons for Africa's rise as a slave producer.* Patrick Manning summarizes the leading theories in chapter 2 of *Slavery and African Life: Occidental, Oriental and African Slave Trades* (New York: Cambridge

University Press, 1990). *Slavery: A World History* by Milton Meltzer, rev. ed. (Cambridge, Mass.: Da Capo Press, 1993) puts slavery in the Americas in a worldwide context. For a review of the slave trade in the Old World, see *Islam's Black Slaves: The Other Black Diaspora* by Ronald Segal (New York: Farrar, Straus & Giroux, 2001).

57   *approximately 12 million enslaved Africans traveled across the Atlantic.* The classic work on the numbers of slaves who crossed the Atlantic is by Philip Curtin, *The Atlantic Slave Trade: A Census* (Madison: University of Wisconsin Press, 1969). Curtin's numbers have been refined several times since then, most notably by the W. E. B. Du Bois Institute in a database of 27,224 slave voyages, available on CD-ROM under the title *The Atlantic Slave Trade, 1527–1867.* A thorough description of the traffic in African slaves appears in *The Slave Trade* by Hugh Thomas (New York: Simon and Schuster, 1997).

58   Map. Estimates of the number of slaves traveling to various parts of the Americas are from Curtin, *The Atlantic Slave Trade,* p. 268. The numbers traveling to Europe and Asia are less well known, though Manning states on p. 171 of *Slavery and African Life* that the number of slaves sent to the Old World was roughly 5 million, compared with the 10 million who reached the New World. He also notes that an additional 15 million people lived as slaves in Africa.

59   *The physical differences among the new occupants of the Americas did little to slow the . . . tendency to interbreed.* Monica Sans describes the genetic mixing of people in Central and South America in "Admixture Studies in Latin America: From the 20th to the 21st Century," *Human Biology* 72 (2000):155– 77. For the situation in Brazil, see *The Brazilians* by Joseph Page (Reading, Mass.: Addison Wesley, 1995). *The Cambridge Encyclopedia of Latin America and the Caribbean* (Cambridge, Eng.: Cambridge University Press, 1985) has a major section on the peoples of Central and South America.

60   *Laws reinforced this distinction.* The legal sanctions against interracial marriages are the topic of David Greenberg's article "White Weddings: The Incredible Staying Power of the Laws Against Interracial Marriage," *Slate* (June 14, 1999).

60   *the Swedish botanist Carolus Linnaeus gave the human species its formal name . . .* Jonathan Marks discusses Linnaeus's system in chapter 3 of *Human Biodiversity: Genes, Race, and History* (New York: Aldine de Gruyter, 1995).

61   *Even before much was known about the genetics of human groups, these arguments were unconvincing.* Arguments against a strict link between genetics and intelligence are presented in chapters 3 through 5 of *Intelligence, Genes, and Success: Scientists Respond to* The Bell Curve, edited by Bernie Devlin, Stephen Fienberg, Daniel Resnick, and Kathryn Roeder (New York: Springer-Verlag, 1997); in chapter 7 of *Human Variation: Races, Types, and Ethnic Groups* by Stephen Molnar, 4th ed. (Upper Saddle River, N.J.: Prentice Hall, 1998); in *The Mismeasure of Man* by Stephen Jay Gould, rev. ed. (New York: W. W. Norton, 1996); and in many other books on human genetics.

63   *The pattern is quite different in other large mammals.* Alan Templeton discusses whether humans can be divided into biological races in "Human

Races: A Genetic and Evolutionary Perspective," *American Anthropologist* 100 (1999):632–50.

63 *If groups can differ in appearance for genetic reasons, they say . . .* See, for example, J. Philippe Rushton, *Race, Evolution and Behavior: A Life History Perspective* (New Brunswick, N.J.: Transaction Books, 1995).

64 *No evidence at all exists that different human groups have ever been under different selective pressures for cognitive traits.* C. Loring Brace makes this point in "An Anthropological Perspective on 'Race' and Intelligence: The Non-Clinal Nature of Human Cognitive Capabilities," *Journal of Anthropological Research* 55 (1999):245–64.

64 *A fundamental distinction exists between a simple trait such as skin color and a complex cognitive attribute such as intelligence.* Ned Block describes this distinction in "How Heritability Misleads about Race," *Cognition* 56 (1995):99–128. A fascinating book on the role of genes in complex human traits is *The Limits and Lies of Human Genetics Research: Dangers for Social Policy* by Jonathan Michael Kaplan (New York: Routledge, 2000).

65 *Physicians have known for a long time that some groups are more prone . . . to particular diseases.* Susan Neuhausen discusses the genetic factors contributing to cancer in "Ethnic Differences in Cancer Risk Resulting from Genetic Variation," *Cancer* 86 (1999):755–62. A discussion of high blood pressure in African Americans can be found in "The Puzzle of Hypertension in African-Americans" by Richard Cooper, Charles Rotimi, and Ryk Ward, *Scientific American* (Feb. 1999):36–43.

65 *group genetic differences also contribute to more common diseases . . .* Two good reviews are "Variations on a Theme: Cataloging Human DNA Sequence Variation" by Francis Collins, Mark Guyer, and Aravinda Chakravarti, *Science* 278 (1997):1580–81; and "DNA Variation and the Future of Human Genetics" by Alan Schafer and J. Ross Hawkins, *Nature Biotechnology* 16 (1998):33–39. For a contrasting point of view, see "Linkage Disequilibrium Mapping of Complex Disease: Fantasy or Reality?" by Joseph Terwilliger and Kenneth Weiss, *Current Opinion in Biotechnology* 9 (1998):578–94.

67 *Dunston and a group of other researchers have been looking at a specific mutation . . .* Dunston's work is described in "Evidence for a BRCA1 Founder Mutation in Families of West African Ancestry" by Heather Mefford, Georgia Dunston, and eight other researchers, *American Journal of Human Genetics* 65 (1999):575–78; and in "BRCA1 Mutations in African Americans" by Ramesh Panguluri, Georgia Dunston, and six other researchers, *Human Genetics* 105 (1999):28–31. For a general review of the work on breast cancer genes, see "Population Genetics of BRCA1 and BRCA2" by Csilla Szabo and Mary-Claire King, *American Journal of Human Genetics* 60 (1997): 1013–20.

4. Encounters with the Other

73 *On May 2, 1932, a team of Palestinian excavators . . .* The leader of the excavation of Skhul, Theodore McCown, describes the discovery in chapter 6 of

The Stone Age of Mount Carmel: Excavations at the Wady el-Mughara, vol. 1, edited by Dorothy Garrod and Dorothea Bate (Oxford, Eng.: Clarendon Press, 1937).

74 *The bones uncovered at Skhul, one of the largest collections of prehistoric human fossils ever found* . . . The fossils from Tabun and Skhul are described in chapter 5 of *In Search of the Neandertals* by Christopher Stringer and Clive Gamble (New York: Thames and Hudson, 1993). Another summary of research on these remains is "Neandertals and Modern Humans in West Asia: A Conference Summary" by Richard G. Klein in *Evolutionary Anthropology* 4 (1995):187–93.

74 *The limb bones were short and thick* . . . Trenton Holliday analyzes the body shapes of the fossils from Skhul and Tabun in "Evolution at the Crossroads: Modern Human Emergence in Western Asia," *American Anthropologist* 102 (2000):54–68.

75 *The fossils from Skhul . . . represented the earliest modern humans ever found outside of Africa.* Ofer Bar-Yosef and Bernard Vandermeersch explain the significance of the fossils from the Middle East in "Modern Humans in the Levant," *Scientific American* (Apr. 1993):94–100.

75 *the two groups encountered each other first in the Middle East* . . . An especially useful article is "The Middle and Early Upper Paleolithic in Southwest Asia and Neighboring Regions" by Ofer Bar-Yosef, in *The Geography of Neandertals and Modern Humans in Europe and the Greater Mediterranean,* edited by Ofer Bar-Yosef and David Pilbeam (Cambridge, Mass.: Peabody Museum of Archaeology and Ethnology, 2000). See also "Modern-Nonmodern Hominid Interactions: A Mediterranean Perspective" by Jean-Jacques Hublin in the same volume.

76 *Two quarrymen were shoveling debris from a limestone cave near Dusseldorf* . . . Erik Trinkaus and Pat Shipman recount the episode at the beginning of *The Neanderthals: Changing the Image of Mankind* (New York: Alfred A. Knopf, 1992).

77 *As the bishop of Worcester's wife famously exclaimed* . . . Alain Corcos recounts this perhaps apocryphal story in *The Myth of Human Races* (East Lansing, Mich.: Michigan State University Press, 1997).

77 *their themes have usually had more to do with the concerns of the day than with established fact.* Paul Graves explores the subtext of the debate in "New Models and Metaphors for the Neandertal Debate," *Current Anthropology* 32 (1991):513–41.

77 *In his 1921 story "The Grisly Folk,"* . . . Wells's story appears in *The Complete Short Stories of H. G. Wells* (London: J. M. Dent, 1998).

78 *one group of paleoanthropologists insisted that Neandertals were the ancestors of today's Europeans.* For a recent exposition of this view, see "Who Were the Neandertals?" by Kate Wong in *Scientific American* (Apr. 2000):99–107.

79 *According to the fossil evidence, the Neandertals evolved from a subbranch of archaic humans* . . . The origins of the Neandertals are described in "Towards a Theory of Modern Human Origins: Geography, Demography, and Diversity in Recent Human Evolution" by Marta Lahr and Robert Foley, *Yearbook of Physical Anthropology* 41 (1998):137–76.

82 *When the paper describing the sequence was published a few months later . . .*
The paper by Krings, Anne Stone, Ralf Schmitz, Heike Krainitzki (the bone
preparer), Mark Stoneking, and Svante Pääbo, "Neandertal DNA Sequences
and the Origin of Modern Humans," appeared in *Cell* 90 (1997):19–30. The
comparison to the Mars landing comes from the news story by Patricia Kahn
and Ann Gibbons, "DNA from an Extinct Human," *Science* 277 (1997): 176–
78.

82 *a result that has been confirmed several more times . . .* Igor Ovchinnikov and
five other researchers describe the results from the second mitochondrial
DNA extracted from a Neandertal in "Molecular Analysis of Neandertal
DNA from the Northern Caucasus," *Nature* 404 (2000):490–93. "Neandertal
Population Genetics" by Matthias Höss, on pp. 453–54 of the same issue of
Nature, explains the significance of the results.

82 *Researchers used this technique to compare chromosomal DNA from a person
living today . . .* This experiment is described in "Genomic Differentiation of
Neandertals and Anatomically Modern Man Allows a Fossil-DNA-Based
Classification of Morphologically Indistinguishable Hominid Bones" by Michael Scholz and seven other researchers, *American Journal of Human Genetics* 66 (2000):1927–32.

83 *According to data gathered by Alfred Kinsey and his colleagues in the 1940s . . .*
See chapter 22 of *Sexual Behavior in the Human Male* by Alfred Kinsey, War-
dell Pomeroy, and Clyde Martin (1948; reprint, Bloomington: Indiana University Press, 1998).

83 *One possibility is that Neandertals and the ancestors of modern humans had
been separated for so long . . .* A forceful advocate of this position is Ian Tattersall. See, for example, "Neandertal Genes: What Do They Mean?" *Evolutionary Anthropology* 6 (1998):157–58.

83 *Some anthropologists have suggested that Neandertal children remained in the
womb for longer than nine months . . .* The idea that Neandertals grew up
more quickly is presented in chapter 4 of *In Search of the Neandertals* by
Stringer and Gamble.

83 *The two separate chimpanzee species that live in equatorial Africa . . .* Morris
Goodman and three coauthors discuss the separation of bonobos and
common chimps in "Primate Phylogeny and Classification Elucidated at the
Molecular Level" in *Evolutionary Theory and Process: Modern Perspectives,* edited by S. P. Wasser (Dordrecht, Netherlands: Kluwer Academic Publishers,
1999).

83 *Perhaps communities of hybrids sprang up over the millennia . . .* For a discus-
sion of this issue see "The Early Upper Paleolithic Human Skeleton from the
Abrigo do Lagar Velho (Portugal) and Modern Human Emergence in Iberia"
by Cidalia Duarte and six coauthors, and the accompanying commentary,
"Hominids and Hybrids: The Place of Neandertals in Human Evolution," by
Ian Tattersall and Jeffrey Schwartz, in *Proceedings of the National Academy of
Sciences* 96 (1999):7117–19.

84 *if hybrids did exist, their contributions to today's gene pool must have been very
small.* Jeffrey Wall declines to draw definitive conclusions about the extent of
genetic mixing in "Detecting Ancient Admixture in Humans Using Sequence

Polymorphism Data," *Genetics* 154 (2000):1271–79. See also "Testing Multi-regionality of Modern Human Origins" by Naoyuki Takahta, Sang-Hee Lee, and Yoko Satta, *Molecular Biology and Evolution* 18 (2001):172–83.

84 *According to one computer model* . . . Ezra Zubrow describes this model in "The Demographic Modelling of Neandertal Extinction," in *The Human Revolution: Behavioural and Biological Perspectives on the Origins of Modern Humans,* edited by Paul Mellars and Christopher Stringer (Princeton: Princeton University Press, 1989).

85 *the moderns at Skhul seem to have behaved almost exactly like the Neandertals* . . . Tal Simmons presents an overview of the interactions between Nean-dertals and modern humans in "Archaic and Modern *Homo sapiens* in the Contact Zones: Evolutionary Schematics and Model Predictions," in *Origins of Anatomically Modern Humans,* edited by Matthew Nitecki and Doris Nitecki (New York: Plenum Press, 1994).

85 *The traditional view is that language arose very early in human history* . . . For an exposition of this view, see Lynne Schepartz, "Language and Modern Hu-man Origins," *Yearbook of Physical Anthropology* 36 (1993):91–126.

85 *When humans give names to things, the idea embodied by that name itself be-comes a tool* . . . See "In the Beginning Was the 'Name'" by Herbert Terrace, *American Psychologist* 40 (1985):1011–28.

87 *Maybe language* . . . *was invented.* This argument is developed at length by William Noble and Iain Davidson in *Human Evolution, Language and Mind: A Psychological and Archaeological Inquiry* (New York: Cambridge University Press, 1996). See also "Neandertals, Modern Humans and the Archaeological Evidence for Language" by Paul Mellars, in *The Origin and Diversification of Language,* edited by Nina Jablonski and Leslie Aiello (San Francisco: Califor-nia Academy of Sciences, 1998) and the special section "The Origins of Speech," *Cambridge Archaeological Journal* 8 (1998):69–94.

89 *One last glimpse at the Neandertals reveals both the tragedy and the inevitabil-ity of their demise.* James Shreeve discusses the Chatelperronian and what he calls the "ultimate private act" of extinction in his book *The Neandertal Enigma: Solving the Mystery of Modern Human Origins* (New York: William Morrow, 1995).

5. Agriculture, Civilization, and the Emergence of Ethnicity

90 *Today little distinguishes Jericho from the other Palestinian towns* . . . Robert Ruby provides a detailed portrait of Jericho's past and present in *Jericho: Dreams, Ruins, Phantoms* (New York: Henry Holt, 1995).

91 *she promoted her work by hinting that she was uncovering proof* . . . Kathleen Kenyon's reflections on the wall and tower of Jericho can be found in *Digging Up Jericho: The Results of the Jericho Excavations 1952–1956* (New York: Praeger, 1957).

93 *They made slender, delicate stone blades.* These tools are described in three papers — "The Origin of Modern Humans" by Ofer Bar-Yosef; "The For-agers of the Upper Paleolithic Period" by Isaac Gilead; and "Complex Hunter/Gatherers at the End of the Paleolithic" by Nigel Goring-Morris — in

The Archaeology of Society in the Holy Land, edited by Thomas E. Levy (New York: Facts on File, 1995). This is a particularly useful book for the Middle East in that it traces the archaeology of the region from the first humans through modern times. See also "The Upper Paleolithic Period in the Levant" by Isaac Gilead, *Journal of World Prehistory* (1991):105–54.

93   *hunter-gatherers tend to organize themselves into groups . . .* In her introduction and section summaries in *Between Bands and States* (Carbondale, Ill.: Center for Archaeological Investigations, 1991), Susan Gregg outlines the history of thinking about bands, tribes, chiefdoms, and states and recent criticisms of those ideas.

94   *Most important, bands can exchange mates.* Robin Fox discusses the genetics of bands in *Kinship and Marriage: An Anthropological Perspective* (1967; reprint, New York: Cambridge University Press, 1983).

94   *bands throughout history have tended to organize themselves into larger units.* For an analysis of this process, see *The Evolution of Human Societies: From Foraging Group to Agrarian State* by Allen Johnson and Timothy Earle, 2nd ed. (Stanford, Calif.: Stanford University Press, 2000).

95   *they would have brought with them a subset of the mitochondrial and Y haplotypes . . .* The haplotypes in northeastern Africa and their distribution in the rest of the world are described in "Different Genetic Components in the Ethiopian Population, Identified by mtDNA and Y-Chromosome Polymorphisms" by Giuseppe Passarino and five other researchers, *American Journal of Human Genetics* 62 (1998):420–34.

96   *this period in Middle Eastern history is called the Natufian.* A good review of the transition to agriculture in the Middle East is "The Natufian Culture in the Levant, Threshold to the Origins of Agriculture" by Ofer Bar-Yosef, *Evolutionary Anthropology* 6 (1998):159–77. Another summary is by Francois Valla, "The First Settled Societies — Natufian (12,500–10,200 BP)" in Levy, ed., *The Archaeology of Society in the Holy Land.*

97   *the wall of Jericho may have had a different purpose.* Ofer Bar-Yosef proposed this idea in "The Walls of Jericho: An Alternative Interpretation," *Current Anthropology* 27 (1986):157–62.

98   *elsewhere other groups of hunter-gatherers also were beginning to experiment with farming.* Many books discuss the transition to agriculture from hunting and gathering. For an overview, see the textbook *People of the Earth: An Introduction to World Prehistory,* 9th ed., by Brian M. Fagan (New York: Longman, 1998). Another good general reference is *The Emergence of Agriculture* by Bruce Smith (New York: W. H. Freeman, 1995). For a summary of recent research, see "The Slow Birth of Agriculture" by Heather Pringle, *Science* 282 (1998):1446–50.

100   *Of the fifty-six wild grasses with the largest seeds, thirty-two grow in the Middle East . . .* Jared Diamond cites this statistic in chapter 8 of *Guns, Germs, and Steel* (New York: W. W. Norton, 1997).

100   *Before the invention of agriculture, the population of the region was probably just a few thousand.* This number comes from the chapter on Asia in *Atlas of World Population History* by Colin McEvedy and Richard Jones (New York: Penguin, 1978).

100 *In a population that is stable or growing slowly . . . women must either limit the number of children . . .* Massimo Livi-Bacci discusses hunter-gatherer populations in chapter 2 of *A Concise History of World Population,* 2nd ed. (Malden, Mass.: Blackwell, 1997).

101 *The resulting growth of population was dramatic.* The number of people on earth at the dawn of agriculture and in A.D. 1 is taken from Livi-Bacci, *A Concise History of World Population.*

101 *How many modern humans have ever lived on the earth?* Kenneth Weiss describes his calculations in "On the Number of Members of the Genus *Homo* Who Have Ever Lived, and Some Evolutionary Implications," *Human Biology* 56 (1984):637–49. I have used his formulas but have recalculated the numbers using the population figures presented in this book. In addition, I've assumed an average lifespan of twenty-two years before the development of agriculture, twenty-five years before the Industrial Revolution, and forty years thereafter.

102 *By 3500 B.C., a massive temple . . . had arisen in the city-state of Uruk . . .* Nicholas Postgate summarizes the history of this period in *Early Mesopotamia: Economy and Society at the Dawn of History* (London: Kegan Paul, 1993).

104 *One way to do this is by fostering the impression that the state is a tribe.* Pierre L. van den Berghe elaborates on this strategy in *Human Family Systems: An Evolutionary View* (New York: Elsevier, 1979).

104 *many of the ancient civilizations in the Middle East were chronically short of labor . . .* William McNeill makes this point in *Polyethnicity and National Unity in World History* (Toronto: Toronto University Press, 1986). For a description of the multicultural aspects of early Middle Eastern civilizations, see "Ethnic Diversity in Ancient Egypt" by Anthony Leahy, in *Civilizations of the Ancient Near East,* vol. 1, edited by Jack Sasson (New York: Scribner, 1995).

105 *These are the groups we know today as ethnic groups.* Thomas Eriksen describes the process by which people come to see themselves as a group in chapter 5, "Ethnicity in History," of *Ethnicity and Nationalism* (London: Pluto Press, 1993). See also the papers "What Is an Ethnic Group: A Biological Perspective" by Helen Macbeth and "What is an Ethnic Group: The View from Social Anthropology" by Ursula Sharma in *Culture, Kinship, and Genes: Towards Cross-Cultural Genetics,* edited by Angus Clarke and Evelyn Parsons (New York: St. Martin's Press, 1997).

105 *Ethnicity is as much a matter of perception as of reality.* For definitions of ethnicity, see chapter 2 of *Ethnicity and Race: Making Identities in a Changing World* by Stephen Cornell and Douglas Hartmann (Thousand Oaks, Calif.: Pine Forge Press, 1998).

6. God's People

106 *The Jews first appear in the historical record as a group of tribes living in the hills around the Dead Sea.* Entire libraries could be filled with volumes written about the history of the Jews and of Judaism. The volumes on which I've relied include *A History of Israel,* 4th ed., by John Bright (Louisville, Ky.:

Westminster John Knox Press, 2000); *Judaism: Between Yesterday and Tomorrow* by Hans Küng (New York: Continuum, 1992); Paul Johnson's *A History of the Jews* (New York: Harper & Row, 1987); *The Illustrated History of the Jewish People*, edited by Nicholas de Lange (New York: Harcourt Brace, 1997); and *The History of Ancient Israel* by Michael Grant (New York: Charles Scribner's Sons, 1984).

107   *Drawing on oral accounts passed down through the generations, they described the origins of their people* . . . Debates over the extent to which the Hebrew Bible can be interpreted as history have become particularly intense in recent years. For an overview, see "Can the Bible Be Trusted?" by Hillel Halkin in *Commentary* 108 (July–August 1999):39–45. A good overview of archaeological work related to the Bible can be found in a series of articles entitled "Archaeology in the Holy Land" by Michael Balter, *Science* 287 (2000):28–35.

107   *The first has been their steadfast belief in a single, all-powerful God* . . . Thomas Cahill explains the significance of this idea for Jewish history — and for world history in general — in *The Gifts of the Jews: How a Tribe of Desert Nomads Changed the Way Everyone Thinks and Feels* (New York: Random House, 1998).

107   *The second has been the creation of a vital ethnic identity* . . . The best exposition I've heard of this idea is in a series of taped lectures by William Scott Green entitled *God and God's People: The Religion of Judaism* (Springfield, Va.: The Teaching Company, 1996).

108   *The researchers called this genetic pattern the Cohen Modal Haplotype.* Karl Skorecki and seven other researchers describe their initial findings in "Y Chromosomes of Jewish Priests," *Nature* 385 (1997):32. See also "Origins of Old Testament Priests" by Mark Thomas and five other researchers, *Nature* 394 (1998):138–39.

109   *most of the Y chromosomes found in Jewish males are the same* . . . The data for the Y chromosomes are from "Jewish and Middle Eastern Non-Jewish Populations Share a Common Pool of Y-Chromosome Biallelic Haplotypes" by Michael Hammer and eleven other researchers, *Proceedings of the National Academy of Sciences* 97 (2000):6769–74. Batsheva Bonné-Tamir and six coauthors discuss early findings on mitochondrial DNA in "Mitochondrial DNA Affinity of Several Jewish Communities," *Human Biology* 65 (1993):359–85.

110   *By A.D. 1 the total Jewish population in the Middle East probably exceeded 5 million.* Sergio DellaPergola discusses the history of Jewish populations in "Major Demographic Trends of World Jewry: The Last Hundred Years," in *Genetic Diversity among Jews: Diseases and Markers at the DNA Level* (New York: Oxford University Press, 1992).

113   *Separate studies of their Y chromosomes and mitochondrial DNA have revealed* . . . See "Mitochondrial DNA Affinity of Several Jewish Communities" by Uzi Ritle and six coauthors, *Human Biology* 65 (1993):359–85; Hammer et al., "Jewish and Middle Eastern Non-Jewish Populations Share a Common Pool of Y-Chromosome Biallelic Haplotypes"; and "The Y Chromosome Pool of Jews as Part of the Genetic Landscape of the Middle East" by Almut

Nebel and five other researchers, *American Journal of Human Genetics* 69 (2001):1095–112.

113 *a team of geneticists decided to take the Lemba's claims more seriously.* The work is described by Mark Thomas and seven other researchers in "Y Chromosomes Traveling South: The Cohen Modal Haplotype and the Origins of the Lemba — The 'Black Jews of Southern Africa,'" *American Journal of Human Genetics* 66 (2000):674–86.

115 *Her first published paper, a 1963 adaptation of her master's thesis . . .* "The Samaritans: A Demographic Study" appeared in *Human Biology* 36 (1963):61–89.

115 *The Samaritans trace their history to the northern kingdom of Israel.* See *The History of the Samaritans*, rev. ed., by Nathan Schur (Frankfurt: Verlag Peter Lang, 1992).

115 *The Samaritans are the most inbred population known anywhere in the world.* Batsheva Bonné-Tamir summarizes her early work on this group in "The Samaritans: A Living Ancient Isolate," in *Population Structure and Genetic Disorders,* edited by Aldur Eriksson (London: Academic Press, 1980).

116 *Some of their members are congenitally deaf . . .* Bonné-Tamir and seven coauthors describe this genetic defect in "Usher Syndrome in the Samaritans: Strengths and Limitations of Using Inbred Isolated Populations to Identify Genes Causing Recessive Disorders," *American Journal of Physical Anthropology* 104 (1997):193–200.

116 *overall rates of fertility are quite similar between groups that marry in and those that marry out.* Alan Bittles and three coauthors describe the consequences of inbreeding in "Reproductive Behavior and Health in Consanguineous Marriages," *Science* 252 (1991):789–94. Other useful reviews are by M. Khlat and M. Khoury, "Inbreeding and Diseases: Demographic, Genetic, and Epidemiologic Perspectives," *Epidemiology Review* 13 (1991):28–41; and by Alan Bittles and James Neel, "The Costs of Human Inbreeding and Their Implications for Variations at the DNA Level," *Nature Genetics* 8 (1994):117–21.

117 *the genetic diseases associated with Jewish populations result from the history of those populations . . .* Arno Motulsky makes this point in "Jewish Diseases and Origins," *Nature Genetics* 9 (1995):99–101.

117 *as a German official put it, "among half-Jews the Jewish genes are notoriously dominant."* Raul Hilberg discusses the ways in which Jews were classified in *The Destruction of the European Jews,* rev. ed. (New York: Holmes and Meier, 1985).

7. The Great Migration

124 *archaeologists exploring the Tarim Basin began to come across ancient cemeteries . . .* A thorough account of the mummies can be found in *The Tarim Mummies: Ancient China and the Mystery of the Earliest Peoples from the West* by James Mallory and Victor Mair (New York: Thames and Hudson, 2000).

127 *Paleoanthropologists have found the remains of an ancient raw bar on the*

beach . . . The shellfish middens near the Horn of Africa are discussed by Robert Walter and eleven coauthors in "Early Human Occupation of the Red Sea Coast of Eritrea During the Last Interglacial," *Nature* 405 (2000):65–69.

127 *A waterborne migration seems likely for another reason.* Christopher Stringer elaborates on the evidence for a coastal route to Southeast Asia in "Coasting Out of Africa," *Nature* 405 (2000):24–26.

128 *Even if modern humans left no archaeological record of their passage, they left a genetic record.* Lluís Quintana-Murci and five other researchers discuss this finding in "Genetic Evidence of an Early Exit of *Homo sapiens sapiens* from Africa Through Eastern Africa," *Nature Genetics* 23 (1999):437–41. For an overview of the hypothesis, see "The Southern Route to Asia" by Todd Disotell, *Current Biology* 9 (1999):R925–28.

129 *Once modern humans reached Australia, they spread quickly . . .* Alan Redd and Mark Stoneking summarize the genetic studies of Australian Aborigines and people from Papua New Guinea in "Peopling of Sahul: mtDNA Variation in Aboriginal Australian and Papua New Guinean Populations," *American Journal of Human Genetics* 65 (1999):808–28.

129 *The remains are clearly modern . . .* Iain Davidson and William Noble describe the early colonization of Australia in "Why the First Colonisation of the Australian Region Is the Earliest Evidence of Modern Human Behaviour," *Archaeology in Oceania* 27 (1992):135–42.

129 *the aborigines have more or less the same genetic variants as all other non-Africans.* For a discussion of these issues using data from chromosome 8, see the paper by Sarah Tishkoff and eleven colleagues, "Short Tandem-Repeat Polymorphism/*Alu* Haplotype Variation at the PLAT Locus: Implications for Modern Human Origins," *American Journal of Human Genetics* 67 (2000):901–25. See also "Independent Histories of Human Y Chromosomes from Melanesia and Australia" by Manfred Kayser and five other researchers, *American Journal of Human Genetics* 68 (2001):173–90.

130 *By about 40,000 years ago, modern humans were living in the interior of southeastern Asia.* Marta Mirazon Lahr and Robert Foley review this colonization in "Multiple Dispersals and Modern Human Origins," *Evolutionary Anthropology* 3 (1994):48–60.

130 *For many years, most paleoanthropologists in China have adhered to a strictly multiregional . . . view . . .* The attitudes of Chinese paleoanthropologists are summarized in "Racial Nationalism and China's External Behavior" by Barry Sautman, *World Affairs* 160, no. 2 (1997):79–95. A broader treatment is *The Discourse of Race in Modern China*, by Frank Dikötter (Stanford, Calif.: Stanford University Press, 1992).

131 *Jin has studied markers on mitochondrial DNA and on the Y and other chromosomes . . .* Evidence for multiple migrations can be found in "Distribution of Haplotypes from a Chromosome 21 Region Distinguishes Multiple Prehistoric Human Migrations," by Li Jin and six colleagues, *Proceedings of the National Academy of Sciences* 96 (1999):3796–800. See also Scott Ballinger and seven coauthors, "Southeast Asian Mitochondrial DNA Analysis Reveals Genetic Continuity of Ancient Mongoloid Migrations," *Genetics* 130 (1992):139–52.

131   *All suggest a gradual population movement from south to north* . . . The genetic
      evidence for south-to-north migration in Asia is summarized in "Natives or
      Immigrants: Modern Human Origin in East Asia" by Li Jin and Bing Su, *Na-
      ture Reviews — Genetics* 1 (2000):126–33. See also "Y-Chromosome Evidence
      for a Northward Migration of Modern Humans into Eastern Asia During the
      Last Ice Age," by Bing Su and twenty other researchers, *American Journal of
      Human Genetics* 65 (1999):1718–24; and "Genetic Relationship of Popula-
      tions in China" by J. Y. Chu and thirteen colleagues, *Proceedings of the Na-
      tional Academy of Sciences* 95 (1998):11763–68.

131   *In China the north is* zhongguo, *the central kingdom, the source of all civiliza-
      tion and culture.* Q. Edward Wang describes this view in "History, Space, and
      Ethnicity: The Chinese Worldview," *Journal of World History* 10 (1999):285–
      305.

131   *most Chinese* . . . *will tell you that they are descended from the Yellow Emperor*
      . . . Lynn Pan recounts the legend of Huang-Ti in *Sons of the Yellow Emperor:
      A History of the Chinese Diaspora* (New York: Kodansha International, 1994).

132   *the Tibetans are descended both from northern Chinese who moved south and
      from central Asian populations.* The origins of the Tibetans are described in
      the paper by Bing Su and ten other researchers, "Y Chromosome Haplotypes
      Reveal Prehistorical Migrations to the Himalayas," in *Human Genetics* 107
      (2000):582–90.

132   *Such population movements have created an incredibly complex mosaic* . . . For
      an example, see "Paternal Population History of East Asia: Sources, Patterns,
      and Microevolutionary Processes" by Tatiana Karafet and eight colleagues,
      American Journal of Human Genetics 69 (2001):615–28.

132   *anthropologists have speculated that these "Mongoloid" characteristics were an
      adaptation to the cold* . . . See, for example, "The Mammoth Steppe and the
      Origin of Mongoloids and Their Dispersal," by R. Dale Guthrie, in *Prehistoric
      Mongoloid Dispersals,* edited by Takeru Akazawa and Emoke Szathmary
      (New York: Oxford University Press, 1996).

133   *The DNA of modern Japanese shows that extensive mixing occurred* . . . Mi-
      chael Hammer and Satoshi Horai investigate the genetics of Japanese popu-
      lations in "Y Chromosomal DNA Variation and the Peopling of Japan,"
      American Journal of Human Genetics 56 (1995):951–62.

134   *The migration of the Yayoi into Japan was part of a much broader movement*
      . . . Takeru Akazawa describes the spread of Asian peoples around the world
      in "Introduction: Human Evolution, Dispersals, and Adaptive Strategies," in
      Prehistoric Mongoloid Dispersals.

135   *Soon after modern people moved into the Middle East to stay, their descendants
      moved northeast* . . . *into Siberia.* The archaeological evidence from Siberia is
      hard to interpret, with a mixture of kinds of fossils and tools. For an over-
      view, see *The Paleolithic of Siberia: New Discoveries and Interpretations,* edited
      by Anatoliy P. Derev'anko (Chicago: University of Illinois Press, 1998). Par-
      ticularly interesting is the chapter by V. Alekseev, "The Physical Specificities
      of Paleolithic Hominids in Siberia."

135   *People and their genes traveled in both directions in northern Eurasia.* The ge-
      netic contribution of Asians to Europeans is described in "Genetic Relation-

ships of Asians and Northern Europeans, Revealed by Y-Chromosomal DNA Analysis," by Tatiana Zerjal and seventeen other researchers, *American Journal of Human Genetics* 60 (1997):1174–83.

136 *The other major place where people from the East met those from the West was in the Tarim Basin.* See "Trading Genes along the Silk Road: mtDNA Sequences and the Origins of Central Asian Populations" by David Comas and eleven other researchers, *American Journal of Human Genetics* 63 (1998):1824–38.

8. Sprung from a Common Source

138 *Yet when these people wanted to refer to the water . . .* John Bengtson and Merritt Ruhlen provide cognates for thirty of these Proto-World words in chapter 14, "Global Etymologies," in Ruhlen's book *On the Origin of Languages: Studies in Linguistic Taxonomy* (Stanford, Calif.: Stanford University Press, 1994). See also chapter 13 in the same book, "The Origin of Language: Retrospective and Prospective."

138 *Most linguists who study the history of language would scoff at this kind of speculation.* Larry Trask provides a clear-eyed assessment of opposing views in the final chapter of *Historical Linguistics* (New York: St. Martin's Press, 1996).

138 *In the King James Bible, which was published in 1611, the Lord's prayer begins this way.* Robert Wright uses this example in his entertaining though caustic article "Quest for the Mother Tongue," *Atlantic Monthly* (Apr. 1991):39–68.

138 *Over 65,000 years . . . the odds of any word staying the same are infinitesimally small.* Many academic linguists make this point in articles and textbooks. See, for example, "The Origin and Dispersal of Languages: Linguistic Evidence" by Johanna Nichols in *The Origin and Diversification of Language,* edited by Nina Jablonski and Leslie Aiello (San Francisco: California Academy of Sciences, 1998).

138 *These similar-sounding words could not have arisen by coincidence, say these linguists . . .* Merritt Ruhlen forcefully makes this argument in *The Origin of Language: Tracing the Evolution of the Mother Tongue* (New York: John Wiley and Sons, 1994).

138 *the correspondences must mean that they descend from a single original language . . .* For a particularly optimistic treatment of Proto-World, see "The Mother Tongue: How Linguists Have Reconstructed the Ancestor of All Living Languages," by Vitaly Shevoroshkin, *Sciences* (May/June 1990):20–27.

139 *the disciplines of archaeology, genetics, and linguistics could be on the verge of a "new synthesis."* Colin Renfrew discussed this idea in "Archaeology, Genetics, and Linguistic Diversity," *Man* 27 (1992):445–78. The need to combine information from multiple disciplines was identified earlier by Luca Cavalli-Sforza in, for example, "Reconstruction of Human Evolution: Bringing Together Genetic, Archaeological, and Linguistic Data," *Proceedings of the National Academy of Sciences* 85 (1988):6002–6.

140 *In 1866 the Linguistic Society of Paris banned all discussions of the origin of language . . .* In their books and articles, many linguists cite this action before launching, as do I, into a discussion of the origins of language.

140   *if two groups speaking the same language are separated for more than about a thousand years . . .* Lyle Campbell provides a thorough description of language change in *Historical Linguistics* (Cambridge, Mass.: MIT Press, 1999).

141   *A remarkable amount is known about the original speakers of . . . Proto-Indo-European.* James Mallory provides a thorough discussion of the origins of these people in *In Search of the Indo-Europeans* (New York: Thames and Hudson, 1989).

142   *The second hypothesis advocates an earlier spread of these languages.* This idea is developed in "The Origins of Indo-European Languages" by Colin Renfrew, *Scientific American* (Oct. 1989):106–14; and "The Early History of Indo-European Languages" by Thomas Gamkrelidze and Vyacheslav Ivanov, *Scientific American* (Mar. 1990):110–16. Renfrew originally proposed the idea in *Archaeology and Language* (Cambridge, Eng.: Cambridge University Press, 1987).

142   *Clear evidence for the domestication of the horse 6,000 years ago is hard to find.* For an excellent overview of this and many other archaeological issues related to historical linguistics, see "At the Edge of Knowability: Towards a Prehistory of Languages" by Colin Renfrew, *Cambridge Archaeological Journal* 10 (2000):7–34.

143   *Today the people of the world speak more than 5,000 languages . . .* The most thorough exposition of the links between languages and language families is in Merritt Ruhlen, *A Guide to the World's Languages* (Stanford, Calif.: Stanford University Press, 1991). An excellent overview of the classification of languages is the 225-page section "Languages of the World" in *Encyclopaedia Britannica.* Another good resource is "The Languages of the World" in *The Cambridge Encyclopedia of Language* by David Crystal (New York: Cambridge University Press, 1987).

143   *any classification will offend some linguists.* An example of some of the controversies that have arisen in attempting to link linguistics and genetics is "Speaking of Forked Tongues: The Feasibility of Reconciling Human Phylogeny and the History of Language," by Richard Bateman and six other authors, *Current Anthropology* 31 (1990):1–24.

144   Map. The distribution of language families is adapted from *The Cambridge Encyclopedia of Language* by David Crystal (New York: Cambridge University Press, 1987), pp. 294–95, and from *Genes, Peoples, and Languages*, Cavalli-Sforza, p. 135.

150   *The distribution of languages does bear some striking resemblances to what is known . . .* Luca Cavalli-Sforza discusses the correspondences among languages, history, and genetics in chapter 5 of *Genes, Peoples, and Languages* (New York: North Point Press, 2000).

150   *various groups of linguists have proposed a connection among language families . . .* Joseph Greenberg made a thorough examination of these connections in *Indo-European and Its Closest Relatives: The Eurasiatic Language Family* (Stanford, Calif.: Stanford University Press, 2000).

151   *The similar-sounding words among languages could be just coincidence . . .* Don Ringe presents the case against reconstruction of ancient languages in

"Language Classification: Scientific and Unscientific Methods," in Sykes, *The Human Inheritance.*

151 *My favorite example is the word "okay."* Many dictionaries of word origins discuss the history of "okay," but the best account I've seen is on the World Wide Web at www.urbanlegends.com/language/etymology/ok_etymology_of.html.

152 *Many of the languages spoken by the aborigines of Australia are quite similar* . . . Robert Dixon discusses the Australian example and many other important issues in *The Rise and Fall of Languages* (New York: Cambridge University Press, 1997).

153 *Where does this leave the "new synthesis" . . . ?* Many questions related to the possibility of a link between languages and genes are covered in the two-volume work *Time Depth in Historical Linguistics,* edited by Colin Renfrew, April McMahon, and Larry Trask (Cambridge, Eng.: McDonald Institute for Archaeological Research, 2000).

154 *thousands of languages are now on the verge of extinction.* For an update on the effort to preserve languages, see "Learning the World's Languages — Before They Vanish" by Bernice Wuethrich, *Science* 288 (2000):1156–59. Daniel Nettle and Suzanne Romaine lay out the case for preserving endangered languages in *Vanishing Voices: The Extinction of the World's Languages* (New York: Oxford University Press, 2000).

9. Who Are the Europeans?

157 *It's Stonehenge, the most famous prehistoric monument in all of Europe.* Good books on Stonehenge include *Stonehenge Complete* by Christopher Chippendale (London: Thames and Hudson, 1994); *The Stonehenge People* by Rodney Castleden (New York: Routledge & Kegan Paul, 1987); and the five chapters on Stonehenge in Aubrey Burl's *Great Stone Circles* (New Haven: Yale University Press, 1999). For a general history of the period, see Tim Darvill, *Prehistoric Britain* (New Haven: Yale University Press, 1987).

158 *The monument was built thousands of years before the druids existed.* The most recent dating is described in *Stonehenge in Its Landscape: Twentieth-Century Excavations* by Rosamund Cleal, K. E. Walker, and R. Montague (London: English Heritage, 1995). A good guide, available at the monument, is *Stonehenge and Neighbouring Monuments* (London: English Heritage, 1995).

159 *Almost all of the mitochondrial and Y-chromosome haplotypes found among Europeans today* . . . For a particularly thorough analysis of European mitochondrial DNA, see "Tracing European Founder Lineages in the Near Eastern mtDNA Pool" by Martin Richards and thirty-six other researchers, *American Journal of Human Genetics* 67 (2000):1251–76.

159 *The Neandertals had a culture well suited to the harsh climate of that time.* Richard Klein summarizes the characteristics of the European predecessors of modern humans in chapter 6, "Neandertals and Their Contemporaries," in *The Human Career: Human Biological and Cultural Origins,* 2nd ed. (Chicago: University of Chicago Press, 1999).

159 *But their culture was no match for that of the newcomers from the Middle East.* A number of papers on the replacement of Neandertals by modern humans in Europe appear in *The Human Revolution: Behavioural and Biological Perspectives on the Origins of Modern Humans,* edited by Paul Mellars and Christopher Stringer (Princeton: Princeton University Press, 1989); and in *The Emergence of Modern Humans: An Archaeological Perspective,* edited by Paul Mellars (Ithaca, N.Y.: Cornell University Press, 1990).

159 *Modern humans . . . brought with them . . . a way of life more sophisticated . . .* Paul Mellars describes the culture of the early modern humans of Europe in chapter 2, "The Upper Palaeolithic Revolution," in *The Oxford Illustrated Prehistory of Europe,* edited by Barry Cunliffe (New York: Oxford University Press, 1994). Chapter 1 in the same book, "The Peopling of Europe 700,000–40,000 Years Before the Present," by Clive Gamble, reconstructs the setting in which modern humans entered Europe. *Prehistoric Europe* by Tim Champion, Clive Gamble, Stephen Shennan, and Alasdair Whittle (London: Academic Press, 1984) gives a broad overview of the period.

159 *They also created art . . .* Pictures of some of the most important artwork in Europe appear in "People Like Us" by Rick Gore, *National Geographic* (July 2000):91–117, one of many articles on European prehistory that have appeared in the magazine.

160 *As the European explorers of the sixteenth and seventeenth centuries encountered people with different appearances . . .* Joseph Graves, Jr., summarizes this history in chapter 3 of *The Emperor's New Clothes: Biological Theories of Race at the Millennium* (New Brunswick, N.J.: Rutgers University Press, 2001).

160 *Consider the people of India.* Genetic studies comparing people in India and Europe are discussed in "An Indian Ancestry: A Key for Understanding Human Diversity in Europe and Beyond" by Toomas Kivisild and nineteen other researchers, in *Archaeogenetics: DNA and the Population Prehistory of Europe,* edited by Colin Renfrew and Katie Boyle (Cambridge, Eng.: McDonald Institute for Archaeological Research, 2000).

161 *Interactions among these groups . . . could then drive social change.* Barry Cunliffe makes this point in his introduction to *The Oxford Illustrated Prehistory of Europe.*

161 *Another factor that contributed to the cultural efflorescence of Stone Age Europe was the climate.* For a climatic overview, see "Where Received Wisdom Fails: The Mid-Palaeolithic and Early Neolithic Climates" by Tjeerd H. van Andel, in Renfrew and Boyle, eds. *Archaeogenetics.*

163 *The conditions that had led to the flourishing of Stone Age culture in Europe were gone.* Paul Mellars describes this period well in chapter 2, "The Upper Palaeolithic Revolution," in Cunliffe, ed., *The Oxford Illustrated Prehistory of Europe.*

163 *some scholars have another hypothesis about the origins of Europeans.* Luca Cavalli-Sforza and Albert Ammerman give the most thorough explanation of their hypothesis in *The Neolithic Transition and the Genetics of Populations in Europe* (Princeton: Princeton University Press, 1984).

164 *These data were ideal, Cavalli-Sforza realized, for studying a particularly contentious issue . . .* Cavalli-Sforza, writing with his son Francesco, describes his

early research on human genetics in chapter 4 of *The Great Human Diasporas: The History of Diversity and Evolution* (Reading, Mass.: Perseus Books, 1995). See also his article "'Genetic Drift' in an Italian Population," *Scientific American* (Aug. 1969):30–37.

166 *Almost immediately he began to find striking patterns of genetic variation.* Cavalli-Sforza, Paolo Menozzi, and Alberto Piazza summarize more than three decades of work on classical genetic markers in *The History and Geography of Human Genes* (Princeton: Princeton University Press, 1994).

166 *Working with the archaeologist Albert Ammerman, he developed a hypothesis.* Ammerman and Cavalli-Sforza first presented their view of the spread of farmers into Europe in "A Population Model for the Diffusion of Early Farming in Europe," in *The Explanation of Culture Change,* edited by Colin Renfrew (London: Duckworth, 1973). Cavalli-Sforza, Paolo Menozzi, and Alberto Piazza describe the underlying concept in "Demic Expansions and Evolution," *Science* 259 (1993):639–46.

167 *The idea was based on "untenable assumptions" and "erroneous interpretations."* The phrases come from a review by Marek Zvelebil of Cavalli-Sforza and Ammerman's book *The Neolithic Transition and the Genetics of Populations in Europe,* in *Journal of Archaeological Science* 13 (1986):93–95. An exchange between Zvelebil and Ammerman appears in three issues of *Antiquity:* 62 (1988):574–83; 63 (1989):162–65; and 63:379–83.

167 *What they have discovered is a process much more complex than was previously recognized.* I. J. Thorpe summarizes the archaeological evidence in *The Origins of Agriculture in Europe* (London: Routledge, 1996). Other books that deal with the introduction of agriculture to Europe include *Europe in the Neolithic: The Creation of New Worlds* by Alasdair Whittle (Cambridge, Eng.: Cambridge University Press, 1996); *Hunters in Transition,* edited by Marek Zvelebil (Cambridge, Eng.: Cambridge University Press, 1986); and chapter 10 of *People of the Earth: An Introduction to World Prehistory,* 9th ed., by Brian Fagan (New York: Longman, 1998).

168 Map. The agricultural frontier in Europe is adapted from *The Significance of Monuments: On the Shaping of Human Experience in Neolithic and Bronze Age Europe* by Richard Bradley (New York: Routledge, 1998), p. 12.

169 *For nearly a thousand years a clear frontier existed.* Richard Bradley analyzes this period and the social dimensions of the monuments in northwestern Europe in *The Significance of Monuments* (New York: Routledge, 1998). Similar analyses appear in Whittle, *Europe in the Neolithic;* in Julian Thomas's *Understanding the Neolithic* (New York: Routledge, 1999); and in "Instruments of Conversion? The Role of Megaliths in the Mesolithic/Neolithic Transition in Northwest Europe" by Andrew Sherratt, *Oxford Journal of Archaeology* 14 (1995):245–60.

172 *He and his colleagues have identified specific mutations of the Y chromosome . . .* The work by Cavalli-Sforza's lab on the Y chromosome is described in "The Genetic Legacy of Paleolithic *Homo sapiens* in Extant Europeans: A Y Chromosome Perspective," by Ornella Semino and sixteen other researchers, *Science* 290 (2000):1155–59. For a news article about the paper, see "Europeans Trace Ancestry to Paleolithic People" by Ann Gibbons in the same issue

of *Science*. Another valuable article is "Y-Chromosomal Diversity in Europe Is Clinal and Influenced Primarily by Geography, Rather Than by Language," by Zoë Rosser and sixty-two other researchers, *American Journal of Human Genetics* 67 (2000):1526–43. In that issue see also "The Peopling of Europe from the Maternal and Paternal Perspectives" by Jeffrey Lell and Douglas Wallace. A general review of the Y chromosome is "New Uses for New Haplotypes: The Human Y Chromosome, Disease and Selection" by Mark Jobling and Chris Tyler-Smith, *Trends in Genetics* 16 (2000):356–62.

172 *Studies of European mitochondria have produced very similar results.* Martin Richards and thirty-six other researchers present a thorough analysis in "Tracing European Founder Lineages in the Near Eastern mtDNA Pool," *American Journal of Human Genetics* 67 (2000):1251–76. Other useful articles on this subject are by Lucia Simoni and four coauthors, "Geographic Patterns of mtDNA Diversity in Europe," *American Journal of Human Genetics* 66 (2000):262–78; Bryan Sykes, "The Molecular Genetics of European Ancestry," *Philosophical Transactions of the Royal Society of London, Biological Sciences* 354 (1999):131–40; and David Comas and five other authors, "Mitochondrial DNA Variation and the Origin of the Europeans," *Human Genetics* 99 (1997):443–49.

174 *the historical processes that led to this extensive degree of genetic mixing . . .* Chapters 9 through 13 of Cunliffe, ed., *The Oxford Illustrated Prehistory of Europe,* provide an overview of these processes.

10. Immigration and the Future of Europe

175 *Immigrants and their children now account for about 9 million of France's 60 million people.* Alec Hargreaves presents a detailed study of the subject in *Immigration, "Race," and Ethnicity in Contemporary France* (London: Routledge, 1995).

176 *just five years ago this suburb and its neighbors were the scene of . . . looting and rioting.* Jonathan Fenby describes this episode in chapter 7 of *France on the Brink: A Great Civilization Faces the New Century* (New York: Arcade, 1998). Chapter 8 of Fenby's book discusses the rise of the National Front.

178 *In polls more than half the French say that too many Muslims are in the country.* See "Extreme Right-Wing Views Rising in France," *Reuters,* May 29, 2000.

178 *every European country has a different relationship with immigration.* For an analysis of the issues underlying immigration policies in Europe, see *Migrants and Citizens* by Rey Koslowski (Ithaca, N.Y.: Cornell University Press, 2000).

179 *Scientists have devoted entire lifetimes to cataloging and measuring how human beings differ.* Joseph Graves, Jr., provides a good overview of this history in *The Emperor's New Clothes: Biological Theories of Race at the Millennium* (New Brunswick, N.J.: Rutgers University Press, 2001). Other valuable sources of information include *Race: History of an Idea in the West* by Ivan Hannaford (Baltimore: Johns Hopkins University Press, 1996); and *On Human Diversity* by Tzvetan Todorov (Cambridge, Mass.: Harvard University Press, 1993).

179 *They contend that Europeans invented racism in the eighteenth and nineteenth centuries . . .* Audrey Smedley makes this claim in *Race in North America: Origin and Evolution of a Worldview,* 2nd ed. (Boulder, Col.: Westview Press, 1999). Dinesh D'Souza largely agrees with her in *The End of Racism* (New York: Free Press, 1995).

180 *They suggest that evolution selected for what has been termed "ethnocentrism."* All sides of this issue are discussed in *Evolutionary Theory and Ethnic Conflict,* edited by Patrick James and David Goetze (Westport, Conn.: Praeger, 2001).

180 *Faith in biblical authority began to break down in the early 1800s.* Audrey Smedley recounts this history in chapter 10 of *Race in North America.*

181 *Ernst Haeckel . . . wrote in his 1905 book . . .* Haeckel is quoted in chapter 5 of *Race and Human Evolution* by Milford Wolpoff and Rachel Caspari (New York: Simon & Schuster, 1997).

181 *the desire to find such differences fueled a mania for human measurement.* Stephen Jay Gould discusses these various measurements in chapter 3, "Measuring Heads: Paul Broca and the Heyday of Craniology," in *The Mismeasure of Man,* rev. ed. (New York: W. W. Norton, 1996). See also "How 'Caucasoids' Got Such Big Crania and Why They Shrank: From Morton to Rushton" by Leonard Lieberman, *Current Anthropology* 42 (2001):69 95.

182 *the cranial index played a prominent role in some of the most bizarre racial theories . . .* The cranial index notion was most fully developed by William Ripley in *The Races of Europe* (New York: Appleton, 1899).

183 *In Germany this theory took a somewhat different form . . .* For a summary of the Aryan ideology, see the epilogue to *In Search of the Indo-Europeans: Language, Archaeology, and Myth* by James Mallory (London: Thames and Hudson, 1989).

183 *All such theorizing could be dismissed as harmless academic sophistry if not for the political potency of these ideas.* Henry Friedlander traces the route from science to the Holocaust in *The Origins of Nazi Genocide* (Chapel Hill: University of North Carolina Press, 1995). See also *Toward the Final Solution* by George Mosse (New York: Howard Fertig, 1978).

183 *Rome's first contact with the rest of Europe came during the great Celtic expansions.* Barry Cunliffe describes the history of the Celts in chapter 10, "Iron Age Societies in Western Europe and Beyond, 800–140 B.C.," in *The Oxford Illustrated Prehistory of Europe* (New York: Oxford University Press, 1994).

184 *in the first century B.C. the Han dynasty in China destroyed the empire of the Xiongnu . . .* Norman Davies recounts the history of the Huns in chapter 4 of *Europe: A History* (New York: Oxford University Press, 1996).

185 *The populations of Europe and Asia . . . were growing.* Population estimates come from *Atlas of World Population History* by Colin McEvedy and Richard Jones (New York: Penguin, 1978).

185 *Today the French see themselves as a single people.* The genetic history of France is described in chapter 5, "Europe," of *The History and Geography of Human Genes* by Luca Cavalli-Sforza, Paolo Menozzi, and Alberto Piazza, abridged ed. (Princeton: Princeton University Press, 1996).

186 Map. The first principal component of genetic variation in France is adapted

from *The History and Geography of Human Genes* by Luca Cavalli-Sforza, Paolo Menozzi, and Alberto Piazza, abridged ed. (Princeton: Princeton University Press, 1994), p. 282.

187 *Suddenly France's birthrates began to fall.* The size of the French population is estimated in *The Cambridge Illustrated History of France* by Colin Jones (New York: Cambridge University Press, 1994). A more general treatment is in *The Demographic Transition: Stages, Patterns, and Economic Implications* by Jean-Claude Chesnais (New York: Oxford University Press, 1992).

188 *France has been relatively open to immigration . . .* See Hargreaves, *Immigration, "Race," and Ethnicity in Contemporary France.*

188 *In most European countries birthrates are now so low that the population cannot replace itself.* Nicholas Eberstadt discusses the upcoming demographic crisis in Europe in "The Population Implosion," *Foreign Policy* (Mar./Apr. 2001):42–53.

188 *Without continued immigration, the population of Europe will begin to decline . . .* The population projections for France and other European countries under different immigration scenarios come from the report *Replacement Migration: Is It a Solution to Declining and Ageing Populations?* by the United Nations Population Division (New York: United Nations Population Division, 2000).

189 *It remains to be seen whether these new people will blend in with local populations . . .* Christopher Caldwell captures many of these issues in "The Crescent and the Tricolor," *Atlantic Monthly* (Nov. 2000):20–34. See also "Questions of Color: Racism in France Persists Despite Egalitarian Creed," *Washington Post,* June 11, 2000:A25.

190 *Already, 45 percent of Algerian men marry French women.* This statistic comes from *Report of the Seminar on Immigration and Integration: Focus on Lyon, France, May 6–8, 1999,* prepared by Migration Dialogue. (Available from Migration News, University of California at Davis.)

11. The Settlement of the Americas

193 *An Indian is an Indian.* C. Loring Brace and A. Russell Nelson discuss the morphological variability of Native Americans in "The Peopling of the Americas: Anglo Stereotypes and Native American Realities," *General Anthropology* 5 (Spring 1999):1–7.

194 *Kennewick Man . . . was obviously a very early inhabitant of the Americas.* Scott Malcomson wrote about Kennewick man in "The Color of Bones," *New York Times Magazine,* Apr. 2, 2000:40–45. Two other articles describing the controversy are by John Miller, "Bones of Contention," *Reason,* Oct. 1, 1997:52–54; and by Steve Coll, "The Body in Question," *Washington Post Magazine,* June 3, 2001:10–13, 21–25.

194 *No human bones more than about 13,500 years old have been found in either North or South America.* Sasha Nemecek presents an overview of the settlement of the Americas in "Who Were the First Americans?" *Scientific American* (Sept. 2000):80–87.

194 *many of the earliest occupants of the Americas didn't look much like current*

Native Americans. Marta Lahr discusses the morphological features of early Americans in "History in the Bones," *Evolutionary Anthropology* 6 (1997):2–6.

195   *In September 2000, the Corps' parent agency, the Department of the Interior, sided with the tribes.* For a news story, see "Bones Decision Rattles Researchers," *Science* 289 (2000):2257.

195   *In 1589 the Jesuit scholar José de Acosta . . .* Sasha Nemecek mentions de Acosta in "Who Were the First Americans?"

195   *Perhaps 75 million people were living in North and South America when Columbus reached the New World . . .* This estimate comes from the entry "Population: Precontact to Present" by Russell Thornton, in *Encyclopedia of North American Indians* (Boston: Houghton Mifflin, 1996).

196   *the only certainty is that people were living in the Americas by about 13,500 years ago . . .* An excellent review of the controversy over when humans first arrived in the Americas is David Meltzer's "Clocking the First Americans," *Annual Review of Anthropology* 24 (1995):21–45.

196   *many of the sites . . . have produced a particular kind of stone tool . . .* National *Geographic* has published a number of articles on the initial occupation of the New World. A recent one is "Hunt for the First Americans" by Michael Parfit, *National Geographic* (Dec. 2000):40–67.

196   *Before about 14,000 years ago, walking south across what is today Canada was not possible . . .* Brian Fagan summarizes the geological evidence in chapter 6, "The First Americans," of *People of the Earth: An Introduction to World Prehistory,* 9th ed. (New York: Longman, 1998).

196   *Within a thousand years or so a corridor had opened up . . .* Michael Wilson and James Burns summarize what is known about this corridor in "Search for the Earliest Canadians: Wide Corridors, Narrow Doorways, Small Windows," in *Ice Age People of North America: Environments, Origins, and Adaptations,* edited by Robson Bonnichsen and Karen Turnmire (Corvallis: Oregon State University Press, 1999).

197   *all of the languages spoken by Native Americans . . . could be grouped into three large families.* Greenberg developed this theory at length in *Language in the Americas* (Stanford, Calif.: Stanford University Press, 1987). A popular account of this proposal appears in "Linguistic Origins of Native Americans" by Joseph Greenberg and Merritt Ruhlen, *Scientific American* (Nov. 1992):94–99.

198   *Christy Turner . . . has been leading an effort to trace the spread of modern humans across the globe using dental evidence.* Turner and G. Richard Scott summarize this work in *The Anthropology of Modern Human Teeth: Dental Morphology and Its Variation in Recent Human Populations* (New York: Cambridge University Press, 1997).

198   *the available genetic evidence also pointed to successive waves of Asian immigrants.* Emöke Szathmary assesses that evidence in "Genetics of Aboriginal North Americans," *Evolutionary Anthropology* 1 (1993):202–20.

198   *Greenberg, Turner, and . . . Stephen Zegura laid out the case for . . . the three-wave model.* See "The Settlement of the Americas: A Comparison of the Linguistic, Dental, and Genetic Evidence," *Current Anthropology* 27 (1986):477–97. A good summary of the three-wave model appears in chapter 6, "Amer-

ica," of *The History and Geography of Human Genes* by Luca Cavalli-Sforza, Paolo Menozzi, and Alberto Piazza (Princeton: Princeton University Press, 1994).

199 *Several sites . . . strongly suggest that people were in the New World well before the appearance of Clovis points.* Robson Bonnichsen and Karen Turnmire survey the status of pre-Clovis sites in *Ice Age People of North America.* A good overview of the pre-Clovis evidence appears in a special issue of *Discovering Archaeology* 2 (Feb. 2000):30–77. See also "Pre-Clovis Sites Fight for Acceptance" by Eliot Marshall, *Science* (Mar. 2, 2001):1730–32.

199 *But the strongest pre-Clovis evidence comes from . . . Monteverde in south-central Chile.* Thomas Dillehay discusses new ideas about the first human occupants of the Americas in *The Settlement of the Americas: A New Prehistory* (New York: Basic Books, 2000).

199 *As one reviewer of Greenberg's proposal put it . . .* This remark, as well as many other criticisms of the three-wave model, appears in the commentaries that accompanied Greenberg, Turner, and Zegura's article in *Current Anthropology.*

200 *The dental evidence has come under similar criticism.* David Meltzer looks at the evidence in "Pleistocene Peopling of the Americas," *Evolutionary Anthropology* 1 (1993):157–69.

200 *Geneticists have been using proteins to study the origins of Native Americans at least since the 1930s.* Michael Crawford summarizes this work in *The Origins of Native Americans: Evidence from Anthropological Genetics* (New York: Cambridge University Press, 1998).

200 *Wallace instead used some of the recently discovered tools of genetic engineering to examine variations in mitochondrial DNA.* Wallace and six colleagues describe their early work with mitochondrial DNA in "Ethnic Variation in *Hpa* I Endonuclease Cleavage Patterns of Human Mitochondrial DNA," *Proceedings of the National Academy of Sciences* 78 (1981):5768–72.

201 *his research has always focused largely on disease.* For a summary of this research, see "Mitochondrial DNA Complex I and III Mutations Associated with Leber's Hereditary Optic Neuropathy" by Michael Brown and five other researchers, *Genetics* 130 (1992):163–72.

201 *People from various parts of the world had distinct patterns of mutations in their DNA.* This work is described in an article by Wallace and four other researchers, "Radiation of Human Mitochondrial DNA Types Analyzed by Restriction Endonuclease Cleavage Patterns," *Journal of Molecular Evolution* 19 (1983):255–71.

201 *many of the conventions used today date from these early investigations.* For an excellent overview of mitochondrial genetics and history, see "Mitochondrial DNA Variation in Human Evolution and Disease" by Doug Wallace, Michael Brown, and Marie Lott, *Gene* 238 (1999):211–30.

202 *Wallace's lab also initiated a much more detailed study of individual haplogroups.* Good summaries of this work appear in "mtDNA Diversity in Chukchi and Siberian Eskimos: Implications for the Genetic History of Ancient Beringia and the Peopling of the New World" by Yelena Starikovskaya and four colleagues, *American Journal of Human Genetics* 63 (1998):1473–91,

and "Mitochondrial DNA and the Peopling of the New World" by Theodore Schurr, *American Scientist* 88 (2000):246–53. See also "Diversity and Age of the Four Major mtDNA Haplogroups, and Their Implications for the Peopling of the New World" by Sandro Bonatto and Francisco Salzano, *American Journal of Human Genetics* 61 (1997):1413–23.

203 *More recent work on the Y chromosome has revealed similarly diverse origins for North American males . . .* See, for example, "Ancestral Asian Source(s) of New World Y-Chromosome Founder Haplotypes" by Tatiana Karafet and thirteen other researchers, *American Journal of Human Genetics* 64 (1999):817–31, and "Y Chromosome Polymorphisms in Native American and Siberian Populations" by Jeffrey Lell and eight colleagues, *Human Genetics* 100 (1997):536–43.

203 *Wallace's lab began to find an unusual Native American haplogroup . . .* For more on haplogroup X, see "mtDNA Haplogroup X: An Ancient Link Between Europe/Western Asia and North America?" by Michael Brown and eight colleagues, *American Journal of Human Genetics* 63 (1998):1852–61.

203 *The settlement of the Americas has always attracted bizarre archaeological hypotheses.* The book *Fantastic Archaeology* by Stephen Williams (Philadelphia: University of Pennsylvania Press, 1991) describes many of these ideas.

204 *But another intriguing possibility remains.* For more on how haplogroup X might have reached the Americas, see "The Solutrean Solution" by Dennis Stanford and Bruce Bradley, *Discovering Archaeology* (Jan./Feb. 2000):54–55.

205 *The European features of skulls such as Kennewick Man are hard to explain . . .* Richard Jantz and Douglas Owsley analyze ancient and modern skulls of Native Americans in "Variation among Early North American Crania," *American Journal of Physical Anthropology* 114 (2001):146–55.

206 *In the 1960 U.S. census — the first that allowed people to classify themselves by racial category . . .* For more on the self-identification of Native American ancestry in U.S. censuses, see Part I of *One Drop of Blood: The American Misadventure of Race* by Scott Malcomson (New York: Farrar, Straus, & Giroux, 2000).

207 *Rather, more people have chosen to recognize or claim some degree of Native American ancestry.* Terry Wilson discusses this trend in "Blood Quantum: Native American Mixed Bloods," chapter 9 in *Racially Mixed People in America,* edited by Maria Root (Newbury Park, Calif.: Sage Publications, 1992).

12. The Burden of Knowledge

208 *Luca Cavalli-Sforza realized that the pieces were falling into place to achieve a lifelong dream.* The article that proposed what would come to be called the Human Genome Diversity Project was by Luca Cavalli-Sforza, Allan Wilson, Charles Cantor, Robert Cook-Deegan, and Mary-Claire King, *Genomics* 11 (1991):490–91.

210 *Cavalli-Sforza . . . had always considered himself a leader in the struggle against racism.* For a summary of Cavalli-Sforza's work on genetics and history, see the book he wrote with his son, Francesco Cavalli-Sforza, *The Great Human*

Diasporas: The History of Diversity and Evolution (Reading, Mass.: Perseus Books, 1995).

210 *a consortium of governments and private organizations had launched the Human Genome Project.* Robert Cook-Deegan, a coauthor of the paper calling for the Human Genome Diversity Project, described the establishment of the Human Genome Project in *The Gene Wars* (New York: W. W. Norton, 1994).

210 *To some critics, sequencing the genome was like "viewing a painting through a microscope."* Kevin Davies includes this remark from the Harvard geneticist Bernard Davis in chapter 1 of *Cracking the Genome: Inside the Race to Unlock Human DNA* (New York: Free Press, 2001).

210 *A report from the National Research Council of the National Academy of Sciences . . .* That report is *Mapping and Sequencing the Human Genome* by the Committee on Mapping and Sequencing the Human Genome (Washington, D.C.: National Academy Press, 1988).

211 *And according to the project's planning document . . .* The "Summary Document" describing the Human Genome Diversity Project is available at the project's Web site: www.stanford.edu/group/morrinst/hgdp/summary93.html.

211 *In 1993 an odd-looking document appeared on the desks of the HGDP's organizers.* RAFI marshals its arguments against the HGDP in "Patents, Indigenous Peoples, and Human Genetic Diversity," RAFI communiqué (May 1993), available on the RAFI home page (rafi.org).

212 *The actions of the U.S. government . . . seemed to bear out RAFI's concerns.* A special issue of the journal *Cultural Survival Quarterly* 20 (Summer 1996) addresses many of the issues raised by the U.S. government's efforts to patent cell lines from indigenous peoples.

213 *The second objection . . . concerns public perceptions of groups.* A committee of the National Research Council reviewed the merits and potential problems of the HGDP in *Evaluating Human Genetic Diversity* (Washington, D.C.: National Academy Press, 1997). Another useful document is "The Human Genome Diversity Project: The Model Ethical Protocol as a Guide to Researchers," a report by Mark Frankel and Antonia Herzog on a symposium and workshop held at the annual meeting for the American Association for the Advancement of Science in February 1998. (Available from the American Association for the Advancement of Science, Washington, D.C.) See also "Is the Human Genome Diversity Project a Racist Enterprise?" by John Moore, in *Race and Other Misadventures: Essays in Honor of Ashley Montagu in His Ninetieth Year,* edited by Larry Reynolds and Leonard Lieberman (Dix Hills, N.Y.: General Hall, 1996).

214 *Greely, who had written widely about the ethical dimensions of genetics . . .* Greely summarizes this work in "Legal, Ethical, and Social Issues in Human Genome Research," *Annual Review in Anthropology* 27 (1998):473–502.

215 *geneticists developed tests that made it possible to identify carriers of sickle cell mutations.* See *In the Name of Eugenics* by Daniel Kevles (Cambridge, Mass.: Harvard University Press, 1995).

216 *he and a group of other ethicists and geneticists put together . . . the "Model Eth-*

ical Protocol." A discussion version of the protocol was published as "Proposed Model Ethical Protocol for Collecting DNA Samples," *Houston Law Review* 33 (1997):1431–73.

217   *But in practice, group consent is not so simple.* E. Juengst describes many of the problems that arise in "Group Identity and Human Diversity: Keeping Biology Straight from Culture," *American Journal of Human Genetics* 63 (1998):673–77.

217   *However, other groups . . . do have recognized governmental structures.* Hank Greely provides an update on the Human Genome Diversity Project, including the reception given the idea of group consent, in "Human Genome Diversity: What about the Other Human Genome Project?" *Nature Reviews — Genetics* 2 (2001):222–27.

218   *the whole process of obtaining consent . . . can seem strikingly one-sided.* Michael Burgess discusses these issues in "Beyond Consent: Ethical and Social Issues in Genetic Testing," *Nature Reviews — Genetics* 2 (2001):147–51.

218   *Some ethicists ask whether informed consent is an unattainable ideal.* See, for example, "Is Non-Directive Genetic Counselling Possible?" by Angus Clarke, *Lancet* 338 (1991):998–1001.

218   *Foster is an advocate of what he calls community review.* Foster and his colleagues describe their work in several related articles: "A Model Agreement for Genetic Research in Socially Identifiable Populations," *American Journal of Human Genetics* 63 (1998):696–702; "The Role of Community Review in Evaluating the Risks of Human Genetic Variation Research," *American Journal of Human Genetics* 64 (1999):1719–27; and "Involving Study Populations in the Review of Genetic Research," *Journal of Law, Medicine & Ethics* 28 (2000):41–51. A particularly useful overview is "Genetic Research and Culturally Specific Risk: One Size Does Not Fit All" by Morris Foster and Richard Sharp, *Trends in Genetics* 16 (2000):93–95.

220   *Government agencies and private companies are compiling huge databases describing the genetic variants . . .* The paper "Haplotype Variation and Linkage Disequilibrium in 313 Human Genes," *Science* 295 (2001):489–93, by J. Claiborne Stephens and twenty-seven other researchers associated with Genaissance Pharmaceuticals in New Haven, Conn., offers a particularly interesting example of the kinds of information being gathered. See also "Drug Firms to Create Public Database of Genetic Mutations" by Eliot Marshall, *Science* 284 (1999):406–7.

13. The End of Race

223   *On the morning of November 26, 1778 . . .* An authoritative biography of Captain Cook is *The Life of Captain James Cook* by John Beaglehole (Stanford, Calif.: Stanford University Press, 1974).

223   *the Polynesians, a people who abided by strict codes of personal hygiene . . .* A good source of information on the daily lives of the Polynesians is the two-volume work by Douglas Oliver, *Oceania: The Native Cultures of Australia and the Pacific Islands* (Honolulu: University of Hawaii Press, 1989).

224   *The young women who swam out to the ships in Hawaii . . .* The entry "French

Polynesia" in *The International Encyclopedia of Sexuality,* edited by Robert Francoeur (New York: Continuum, 1997), contains a wealth of information on the sexual practices of the Polynesians, including Hawaiians, before and after contact with Europeans. See also "Shipboard Relations Between Pacific Island Women and Euroamerican Men 1767–1887" by David Chappell, *Journal of Pacific History* 27 (1992):131–49.

224 *At least 300,000 people, and possibly as many as 800,000, lived on the Hawaiian Islands* . . . The figures are from *Before the Horror: The Population of Hawai'i on the Eve of Western Contact* by David Stannard (Honolulu: Social Science Research Institute, University of Hawai'i at Manoa, 1989).

224 *By the time the painter Paul Gauguin journeyed to the Pacific in 1891* . . . Alan Moorehead describes Gauguin's reactions to Tahiti in *The Fatal Impact* (New York: Harper & Row, 1966).

225 *Almost half the people who live in Hawaii today are of "mixed" ancestry.* Michael Haas discusses the prominence of interethnic marriages in his Introduction to *Multicultural Hawai'i: The Fabric of a Multiethnic Society,* edited by Haas (New York: Garland, 1998).

225 *Hawaii's high rates of intermarriage have fascinated academics for decades.* For a summary of the work of Romanzo Adams and other sociologists, see "The Illusion of Paradise: Privileging Multiculturalism in Hawai'i" by Jonathan Okamura, in *Making Majorities,* edited by Dru Gladney (Stanford, Calif.: Stanford University Press, 1998).

225 *because of kinship ties, American families are already much more mixed than they look.* Joshua Goldstein documents the extent of this interrelatedness in "Kinship Networks That Cross Racial Lines: The Exception or the Rule?" *Demography* 36 (1999):399–407.

227 *In the Yugoslavian war, the Croats caricatured their Serbian opponents* . . . Marek Kohn cites this example in chapter 1 of *The Race Gallery: The Return of Racial Science* (London: Jonathan Cape, 1995).

227 *In Africa the warring Tutsis and Hutus often call attention to the physical differences of their antagonists* . . . Stephen Cornell and Douglas Hartmann discuss ethnicity in Africa in *Ethnicity and Race: Making Identities in a Changing World* (Thousand Oaks, Calif.: Pine Forge Press, 1998).

227 *The Polynesian colonizers of the Hawaiian archipelago are a good example.* A news article describing this research is "The Peopling of the Pacific" by Ann Gibbons, *Science* 291 (2001):1735–37. Good overviews of this topic are "Molecular Genetic Evidence for the Human Settlement of the Pacific: Analysis of Mitochondrial DNA, Y Chromosome, and HLA Markers" by Erika Hagelberg and seven coauthors, *Philosophical Transactions of the Royal Society of London B* 354 (1999):141–52 and "Foregone Conclusions? In Search of 'Papuans' and 'Austronesians'" by John Terrell, Kevin Kelly, and Paul Rainbird, *Current Anthropology* 41 (2001):97–124.

229 *Geneticists Manfred Kayser and Mark Stoneking* . . . *have dubbed the resulting synthesis the "slow-boat model."* Stoneking, Kayser, and five other authors lay out this model in "Melanesian Origin of Polynesian Y Chromosomes," *Current Biology* 10 (2000):1237–46. See also "A Predominantly Indigenous Paternal Heritage for the Austronesian-Speaking People of Insular Southeast

Asia and Oceania" by Cristian Capelli and nine colleagues, *American Journal of Human Genetics* 68 (2001):432–43.

229 *The Polynesians first reached the Hawaiian Islands around* A.D. *400* . . . Peter Bellwood gives an overview of the settlement of Polynesia in "The Austronesian Dispersal and the Origin of Languages," *Scientific American* (July 1991):88–93.

229 *The early decades of the nineteenth century brought just a trickle of settlers to the islands* . . . For a good overview of Hawaii's multiethnic history, see chapter 2, "A Brief History," in Haas, *Multicultural Hawai'i*. Bernhard Hormann describes how groups mixed in "The Mixing Process," *Social Process in Hawaii* 29 (1982):117–29.

229 *Over the next century nearly half a million more workers followed.* Andrew Lind reviewed immigration to Hawaii in "Immigration to Hawai'i," *Social Process in Hawaii* 29 (1982):9–20.

230 *But there's actually a fair amount of prejudice here.* For some written descriptions, see "Ethnic Antagonism and Innovation in Hawaii" by John Kirkpatrick, in *Ethnic Conflict: International Perspectives*, edited by Jerry Boucher, Dan Landis, and Karen Arnold Clark (Newbury Park, Calif.: Sage, 1987); the chapter "Literature" by Rodney Morales (and several other chapters) in Haas, *Multicultural Hawai'i*; "Race and Ethnic Relations: An Overview" by Andrew Lind, *Social Process in Hawaii* 29 (1982):130–50; and Paul Theroux, *The Happy Isles of Oceania: Paddling the Pacific* (New York: G. P. Putnam's Sons, 1992).

231 *Okamura holds that racial and ethnic prejudice is deeply ingrained* . . . Jonathan Okamura presents his views in "The Illusion of Paradise: Privileging Multiculturalism in Hawai'i" in Gladney, *Making Majorities,* and in "Social Stratification" in Haas, *Multicultural Hawai'i.*

232 *Greater recognition of the value and fragility of this culture has led to a resurgence of interest* . . . Jay Hartwell profiles this resurgence in *Na Mamo: Hawaiian People Today* (Honolulu: 'Ai Pohaku Press, 1996).

232 *For the past several decades a sovereignty movement has been building among Native Hawaiians* . . . Two books that bracket the range of views on Native Hawaiian sovereignty are *From a Native Daughter: Colonialism and Sovereignty in Hawai'i,* rev. ed., by Haunani-Kay Trask (Honolulu: University of Hawai'i Press, 1999) and *Hawaiian Sovereignty: Do the Facts Matter?* by Thurston Twigg-Smith (Honolulu: Goodale, 2000). A more balanced perspective on sovereignty issues can be found in *Hawai'i Politics and Government: An American State in a Pacific World* by Richard Pratt with Zachary Smith (Lincoln: University of Nebraska Press, 2000).

233 *"Pure" Hawaiians with no non-Hawaiian ancestors probably number just a few thousand.* Pratt's *Hawai'i Politics and Government* gives population sizes for ethnic groups in Hawaii.

233 *Past legislation has waffled on this issue.* Jon Van Dyke, a professor of law at the University of Hawaii, has written extensively about legislation affecting Native Hawaiians. See, for example, "The Political Status of the Native Hawaiian People," *Yale Law and Policy Review* 17 (1998):95.

233 *No one is better qualified to judge this idea than Rebecca Cann.* Cann provides

an overview of anthropological genetics in "Genetic Clues to Dispersal in Human Populations: Retracing the Past from the Present," *Science* 291 (2000):1742–48. For a sample of her work on Pacific populations, see "mtDNA and Language Support a Common Origin of Micronesians and Polynesians in Island Southeast Asia," *American Journal of Physical Anthropology* 105 (1998):109–19.

234   *But some have mitochondrial DNA from elsewhere — southern Africa, or eastern Asia, or even Polynesia . . .* Bryan Sykes cites these examples in *The Seven Daughters of Eve: The Science That Reveals Our Genetic Ancestry* (New York: W. W. Norton, 2001).

235   *In some populations in South America, virtually all the Y chromosomes are from Europe and all the mitochondrial DNA is from indigenous groups.* L. G. Carvajal-Carmona and ten coauthors document this distribution in "Strong Amerind/White Sex Bias and a Possible Sephardic Contribution among the Founders of a Population in Northwest Colombia," *American Journal of Human Genetics* 67 (2000):1062–66.

235   *30 percent of the Y chromosomes in African-American males come from European ancestors.* Arthur Allen wrote about this project on May 12, 2000, in the on-line magazine *Salon* (http://www.salon.com).

236   *The most remarkable aspect of ethnicity in Hawaii is its loose relation to biology.* David Hollinger discusses people's associations with "ethno-racial blocs" in *Postethnic America: Beyond Multiculturalism* (New York: Basic Books, 1995). Alain Touraine explores related issues in *Can We Live Together: Equality and Difference* (Cambridge, Eng.: Polity Press, 2000).

236   *Hawaii's high rates of intermarriage also contribute greatly to the islands' ethnic flux.* Ronald Johnson explores the implications of intermarriage in "Offspring of Cross-Race and Cross-Ethnic Marriages in Hawaii" in *Racially Mixed People in America,* edited by Maria Root (Newbury Park, Calif.: Sage Publications, 1992). For a broader overview of the issues associated with intermarriage, see *Mixed Blood: Intermarriage and Ethnic Identity in Twentieth-Century America* by Paul Spickard (Madison: University of Wisconsin Press, 1989).

236   *ethnic and even "racial" groups still exist in Hawaii . . .* For a somewhat different perspective on this issue, see *The Future of Ethnicity, Race, and Nationality* by Walter Wallace (Westport, Conn.: Praeger, 1997). See also "Intermarriage and the Future of Races in the United States" by Roger Sanjek, in *Race,* edited by Steven Gregory and Roger Sanjek (New Brunswick, N.J.: Rutgers University Press, 1994).

238   *In his novel* Siddhartha, *Hermann Hesse tells the story of a young man in ancient India . . . Siddhartha* by Hermann Hesse, translated by Hilda Rosner (New York: New Directions, 1951).

Acknowledgments

Writing this book was great fun. I traveled to parts of the world I never imagined I'd see. I spent countless hours in libraries reading about exotic times, events, and places. And I was privileged to talk with some of the most fascinating people I've ever met.

Many geneticists, archaeologists, paleoanthropologists, linguists, and other scholars gave generously of their time to talk with me and to review parts of the manuscript. They include David Altshuler, George Armelagos, Michael Bamshad, Mark Batzer, Batsheva Bonné-Tamir, Alison Brooks, Lisa Brooks, Michael Brown, Rebecca Cann, Luca Cavalli-Sforza, Aravinda Chakravarti, Joseph Chang, Francis Collins, Glenn Conroy, Robert Cook-Deegan, Georgia Dunston, Marcus Feldman, Morris Foster, David Goldstein, Joshua Goldstein, Hank Greely, Colin Groves, Michael Hammer, Edward Hammond, Rosalind Harding, Henry Harpending, Debra Harry, John Hawks, Jody Hey, Trefor Jenkins, Li Jin, Lynn Jorde, Kenneth Kidd, Mary-Claire King, Matthias Krings, Jeffrey Lell, Vincent Macaulay, Sally McBrearty, Jonathan Marks, John Moore, Joanna Mountain, Arno Motulsky, Elizabeth Nickerson, Peter Oefner, Jonathan Okamura, Igor Ovchinnikov, Svante Pääbo, Andrew Pakstis, Jane Phillips-Conroy, Jonathan Pritchard, William Provine, Molly Przeworski, Jennifer Reardon, Colin Renfrew, Merritt Ruhlen, Andres Ruiz-Linares, Ralf Schmitz, Theodore Schurr, Xiufeng Song, Himla Soodyall, Anne Stone, Mark Stoneking, Christopher Stringer, Zejing Tan, Alan Templeton, Joseph Terwilliger, Sarah Tishkoff, Philip Tobias, Antonio Torroni, Chris Tyler-Smith, Peter Underhill, Craig Venter, Douglas Wallace, James Watson, Kenneth Weiss, Mark Weiss, Carsten Wiuf, Bernard Wood, Stephen Wooding, and Yongqin Xu.

Several of these scientists were especially helpful while I was doing research for this book. Luca Cavalli-Sforza at Stanford University sparked my interest in the historical dimensions of genetics. James Watson at the Cold Spring Harbor Laboratories inspired me with his unabashed enthusiasm for anthropological genetics; he also suggested that I go to Botswana to meet with the Bushmen. Lynn Jorde at the University of Utah, Joanna Mountain and Peter Underhill at Stanford University, Sarah Tishkoff at the University of Maryland, and Mark Weiss at the National Science Foundation met and talked with me repeatedly as I struggled through various issues. Stephen Wooding of the University of Utah read through the entire manuscript and provided much-needed help with the science. For their hospitality, I thank Li Jin in Shanghai, Batsheva Bonné-Tamir in Tel Aviv, François and Florence Veyrié in Lyon, Rebecca Cann in Honolulu, and Himla Soodyall in Johannesburg. Jeff Gush and Cherri Briggs of Explore, Inc., who are working hard to better the lives of the Bushmen in southern Africa, helped arrange my travels in Botswana. Thanks, too, to Kathryn Beauregard of the *American Journal of Human Genetics*, where many pioneering articles on genetics and human history have been published.

This book would never have been written if not for the efforts of two people I count as both colleagues and friends. My agent, Rafe Sagalyn, saw the potential in this topic the moment I described it to him and urged me not to be intimidated by such a sprawling story. My editor at Houghton Mifflin, Laura van Dam, also embraced the idea for this book from the very beginning. Later, her structural suggestions and word-by-word editing improved every paragraph of the manuscript.

Also at Houghton Mifflin, Peg Anderson provided insightful and sensitive editing; Dan O'Connell organized publicity; Erica Avery was a knowledgeable and friendly point of contact; and Janet Silver made key decisions at important stages in the writing process. My editor at the *Atlantic Monthly*, Katie Bacon, helped me think through several critical issues. Bonnie Branner, a freelance artist in Washington, D.C., drew the maps. Sidney Richman took the photograph on the jacket.

Many friends and relatives offered invaluable help and encouragement. David Jarmul, David Reiser, and Blake Rodman read an early draft of the manuscript and made many excellent suggestions. Many others read parts of the book or buoyed my spirits with their interest in my work. Thanks especially to Doug Canter, Jon Cohen, Diana Deem, Gregg Easterbrook, Betsy Falloon, Rome Hartman, Patti Hatleberg, Ava Kaufman, Nan Kennelly, Dan Kois, Dave Olson, Lisa Olson, Rick Olson, Adele and Sidney Richman, Jack Shafer, Umberto Vizcaino, and Fred Wine (for the Friday morning basketball games).

Finally, my family was a constant supportive presence, even when my research took me far from home. Lynn Olson and Diane Olson read various versions of chapters and always helped me improve them. And my children, Eric and Sarah, continually reminded me that what I was doing was important.

Index